Finite Element Analysis in Manufacturing Engineering

Other McGraw-Hill Books of Interest

Finite Element Analysis in Manufacturing Engineering

Edward R. Champion, Jr.

McGraw-Hill, Inc.

New York St. Louis San Francisco Auckland Bogotá
Caracas Lisbon London Madrid Mexico Milan
Montreal New Delhi Paris San Juan São Paulo
Singapore Sydney Tokyo Toronto

Library of Congress Cataloging-in-Publication Data

Champion, Edward R.,
    Finite element analysis in manufacturing engineering /
    Edward R. Champion, Jr.
        p.   cm.
    Includes bibliographical references and index.
    ISBN 0-07-010510-3
    1. Finite element method.   2. Production engineering—Mathematics.
    I. Title.
    TA347.F5C45   1992   91-44248
    670.42—dc20   CIP

1 2 3 4 5 6 7 8 9 0   DOC/DOC 9 8 7 6 5 4 3 2

ISBN   0-07-010510-3

The sponsoring editor for this book was Gail Nalven, and the production supervisor was Suzanne W. Babeuf. It was set in Century Schoolbook by North Market Street Graphics.

Printed and bound by R. R. Donnelley & Sons Company.

To My Family

Contents

List of Figures

CHAPTER 8

CHAPTER 9

CHAPTER 10

CHAPTER 11

List of Tables

Preface

This book is intended to be a basic guide for the manufacturing engineer who wants exposure to the use of Finite Element Analysis (FEA) in a small to medium size manufacturing facility where either personnel or hardware (or both) resources are limited. No previous exposure to FEA is assumed. Moreover, this book is applicable to students in the science and engineering fields who wish to obtain a concise introduction to the subject of applied FEA, and who wish to learn how to apply this analysis technique to their particular situation.

In keeping with this theme, the reader will find the text directed toward helping the user solve general classes of engineering problems with FEA on personal computers. The types of problems that may be solved on personal computers are primarily limited by memory, hard disk space, and overall throughput of the system. Problems that previously involved cumbersome or intractable solutions can now be analyzed and solved if the engineer owns or has access to a personal computer and the proper software.

There are advantages and disadvantages in using a personal computer for finite element analysis. Certainly a motivating factor that has fostered the popularity of the PC/FEA combination has been the relatively low cost of this type of analysis. However, there are often restrictions on problem size coupled with long execution times.

With the newer and faster processors that continually

come to market, the availability of massive amounts of storage, and software refinements in algorithms and code that allow the fastest possible execution and minimize space requirements, rather complex problems can be handled. These types of problems range from linear analysis to fluid flow to highly nonlinear analysis.

All problems run in this book, with the exception of one in Chap. 9 that used ANSYS, were executed with the program SuperSAP (ALGOR Interactive Systems, Inc., 260 Alpha Drive, Pittsburgh, PA 15238, 412-967-2700. In addition, all problems were run using an 80386 25 MHz personal computer.

Chapter 1 addresses some of the basic requirements for finite element analysis and what to look for when considering using this tool.

The remainder of the text is divided into the following chapters:

Chapter 2 covers the practical aspects of hardware requirements plus recommendations for upgrades to your existing computer systems. Discussions in this chapter range from the basic PC unit and upgrades to the basic unit to get the most performance possible within a given budget.

Chapter 3 looks at the fundamental aspects of finite element analysis such as setting up problems and element types. There is discussion on selecting proper mesh sizes, elements, etc.

Chapter 4 gives an introduction to the ALGOR Interactive Systems, Inc. finite element program and explains how to use the program with sample verification problems and a detailed example, both from an FEA standpoint and a conventional solution.

Chapter 5 covers the finite element methodology in somewhat more detail and presents a FORTRAN program for calculation of two-dimensional heat transfer.

Chapter 6 offers real-world applications of FEA in a manufacturing environment with applications to tools, molds, and dies.

Chapter 7 illustrates finite element analysis uses in the automotive industry.

Chapter 8 presents a unique view of using FEA in a musical product modification.

Chapter 9 examines finite element analysis as used in some applications in a military environment.

Chapter 10 looks at a more commercial side as it examines applications to in-home medical products.

Chapter 11 applies the finite element concept to situations within the utility industry.

The Reference section contains useful references to common finite element texts and should serve as a stepping stone to further reading on the subject.

All references to specific trade names are copyrighted by the respective companies.

Acknowledgments

For their assistance in the development and writing of this book, I would like to thank the following individuals:

Dr. Joel Orr, the McGraw-Hill Series Editor

Ms. Gail Nalven, Senior Editor at McGraw-Hill

Ms. Christine Furry, North Market Street Graphics

ALGOR Technical Support, Mr. Mark Dekker, and Mr. Pat Cronin

WordPerfect Technical Support Staff

Mr. Harold Lawson, Moore Special Tool Company

Mr. Ken Woodard, Kollsman

Mr. Clark Wilson, Warn Industries

Mr. Marvin Zeigler, Wave Air, Inc.

Mr. Warren Peters, Electri-Glass, Inc.

Mr. Paul Levering, Webb Wheel Products

Mrs. Joyce Champion for the support and equipment

Last but certainly not least, I want to thank my wife, Ginger, and my daughter, Caroline, for putting up with the long office hours required to complete this task, in addition to performing all the activities required to maintain a home.

Finite
Element Analysis in
Manufacturing
Engineering

1

Applied Finite Element Analysis in Manufacturing Engineering

1.1 Introduction

What is Manufacturing Engineering and how is Finite Element Analysis related to such a branch of engineering? Manufacturing Engineering can be thought of as that branch of engineering that synthesizes the design engineer's product ideas and paper (or CAD) designs, and the real-world capability to produce the design with existing tools and/or machinery. Given the above premise, one has to have the tools and insights to a design in order to effectively synthesize that design into a valid or optimized product. The focus of this text is to introduce the reader to the use of Finite Element Analysis (FEA) in order to provide a practical solution to manufacturing engineering problems.

To use FEA intelligently (that is, to be able to model properly, apply the correct boundary conditions, and correctly interpret the results of the analysis), certain basic information is needed. In the remainder of Chap. 1, I discuss briefly the history of FEA and why one would consider using FEA. Chapter 2 discusses the necessary hardware and enhancements that you may want to make to your existing hardware in order to provide an acceptable performance level.

1

Chapter 3 continues with a review of generalized concepts of FEA, and Chap. 4 reviews the software package selected for this text and a simple example of an FEA problem. Chapter 5 discusses the development of a routine to solve a 2-dimensional heat transfer problem and the FORTRAN code is given. The remaining chapters address typical real-world problems and their solutions. The problems are worked out showing each step in the problem definition, application of boundary conditions, application of loading, analysis of the results, and discussion of the results.

1.2   Why FEA?

The question is often asked, "Why should FEA be used to solve my problem(s)?" This is a valid question and one that deserves a rational answer. A preconception of those who do little analysis during a product design is that their existing design techniques have always worked in the past and the existing design techniques should work in the future. While this might have been and still might be the case, little thought is given to the real economics and subsequent optimization of the design. Every bit of material not necessary for the functioning of a particular product adds to the overall product cost in raw material costs, production costs (complexity of machinery required to build the product), shipping costs (on high volume items), and general overhead costs associated with the product. The bottom line for any product is cost and profitability. Today's market, which includes not only local market areas, but also international market areas, is highly competitive. To be competitive, the manufacturer must produce the best product at the lowest cost.

This text is intended as an introduction to using FEA in the manufacturing environment in order to produce the optimum product: a product that performs as intended, meets all of the specified environmental requirements, and is the least costly to produce.

1.3   A Brief History of Finite Element Analysis

This section covers some of the fundamental papers and books in which the reader may wish to explore the basis of the finite element method. This very brief summary highlights papers and books that I have found useful in gaining background and insight into the basis of FEA.

What is the Finite Element Method? The Finite Element Method is a numerical analysis procedure to obtain approximate solutions to problems posed in every field of engineering. Often problems encountered in everyday practical applications do not lend themselves to closed-form solutions. This is readily apparent if, for example, one wishes to analyze the flow of a fluid through a duct of varying cross section subjected to nonuniform boundary conditions. We can write the governing equations with the appropriate initial and boundary conditions. However, it is apparent that no simple analytic or closed-form solution exists. This complication is found not only in the fluid mechanics area, but also in the areas of structural mechanics and heat transfer analysis.

From this inability to solve many complex structural problems arose the foundations of the finite element method as we know it today.

The label *finite element method* appeared in 1960 in a paper by Clough[1] concerning plane elasticity problems. However, the basis for the finite element method goes back further to the early 1940s in the applied mathematics literature. We will not cover that period of development, for there are sufficient reviews available concerning this area of the history of the finite element method. During the mid- to late-1950s, the mathematical literature on the finite element method increased significantly. Books and monographs[2-4] were written to explain the mathematical foundations of the method during the early 1970s. Not only were the mathematicians exploring the finite element method basics, but the physicists were also developing ideas and concepts along similar lines. One such example is the development of the hypercircle method, a concept in function space, that allows a geometric interpretation of minimum principles when associated with the classical theory of elasticity.

During the mid- to late-1950s and early 1960s a series of papers were published covering linear structural analysis and methods to effect efficient solutions to these problems. Turner, Clough, Martin, and Topp[5] derived solutions to plane stress problems via triangular elements with properties determined from elasticity theory. As the digital computer evolved, these solution algorithms allowed the analysis of more complex problems.

Concepts of the method began to be drawn into focus by the publishing of many papers and articles.[6-10] The method was recognized as a form of the Ritz method, and also as a technique to solve elastic continuum problems. In 1965, Zienkiewicz and Cheung[11] illustrated the applicability of the finite element method to any field problem that could be formulated by variational means. Most of the applications of the finite element method during the early- to mid-1970s were in the structural analysis area since the original development came from the structural analysis area. This is the reason why many FEA programs today are heavily oriented toward the structural analysis discipline.

As the structural analysis applications of FEA were refined, other areas of applications of FEA were nurtured during the late 1970s and early 1980s.[6-12] Attention was and is being given to the application of FEA to the thermal/fluids areas. For example, many papers in the mid- to late-1970s applied the finite element method to the Navier-Stokes equations. Papers by Girault and Raviart[12] and Teman[13] have lead to further development of this method in fluids applications. The reader should refer to the works of Fortin,[14] Griffith,[15] Thomasset,[16] Heywood and Rannacher,[17] and Cullen[18] for the basic understanding of the fluids modeling effort. In addition, Chung[19] has an excellent text on FEA in fluid dynamics. If the user wishes a fundamental text concerning FEA in general, I recommend texts by Gallagher,[20] Segerlind,[21] or Zienkiewicz and Taylor.[22]

1.4   Applications of FEA for Practicing Engineers

Every area of engineering can use the power of the Finite Element Method of Analysis. From the history of the analytic technique, the most obvious application is structural analysis. The fields of Civil and Aerospace Engineering rely heavily on FEA methods to analyze various structures ranging from buildings to spacecraft design. The analysis involves the determination of static deflections and stresses and the determination of natural frequencies and modes of vibration. In addition to these types of problems, the ability exists to analyze the stability of structures, fatigue, shock, random vibration, random time-based loading, and even optimization of structures for weight, strength, or an array of factors.

Other FEA applications lie in the thermal/fluids areas. Many models contain thermal and fluid elements which will allow the determination of temperature distributions in structures and such items as velocity, pressure, and concentration distributions in fluid flows. This type of analysis is very important in external atmospheric structure loadings, piping analysis, etc.

There are three basic types of problems, from a mathematical standpoint, that the practicing engineer faces. These are Time Independent (Equilibrium) Problems, Eigenvalue Problems, and Time Dependent (Propagation) Problems. When an engineer is faced with a problem that does not lend itself to an easily tractable solution, he or she has to make judgments as to the best solution approach in light of his background and familiarity with the tools available. The available tools include, but are not limited to, computer resources and programs to utilize the computer resources. Unfortunately, not every engineer has at his or her disposal a large mainframe computer or a personal desktop computer, the finite element code, and the time to solve whatever problems arise. Consequently, the engineer is forced to make approximations based on similarity to classical, closed-form soluble problems, or the engineer may rely on physical prototyping. Physical prototyping is generally expensive and increases the cost of the design process. These approximations may or may not result in an adequate design.

An overdesigned (or overengineered) product may be noncompetitive from a cost standpoint and/or may actually not perform as intended. However, this design process is often tempered by experience. With the proliferation of personal computers and the availability of software to aid the design process, it is no longer necessary to rely heavily on intuition and settle for approximate solutions to real-world problems. FEA programs are available to aid the design engineer in developing a product to withstand its environment.

The better FEA products interface with a variety of CAD programs. The interface allows a rapid design and analysis cycle with many *"what if"* scenarios.

1.5 Using This Book

If the user/engineer is not as experienced as he or she wishes to be, then this book can aid in the development of a better foundation of understanding in the basics of analysis through worked examples. The application of FEA to a wide variety of problems that may be encountered in the day-to-day engineering activities can be readily handled. The following section addresses how one may use this book as an aid and guide to FEA with personal computers. Since the basic objective of this text is to familiarize the novice as well as the experienced user with tools that are available and necessary to analyze and solve problems with Finite Element Methods and the application of these tools, I suggest the following two approaches in the use of this book.

1.5.1 The novice or casual user

If you are a part of the novice or casual-user category, the first question to be answered is, "Do I have an application or problem that requires the use of Finite Element techniques?" I recommend that you review Chap. 3. Section 3.2 addresses the criteria for selecting the finite element approach to solving a problem. Perhaps there are alternate solution techniques which you have overlooked. However, if you are convinced FEA is the direction you wish to pursue, then continue in Chap. 3 and develop a familiarity with the basic concepts and terms of FEA. The next step is to evaluate the equipment requirements in order to perform an analysis. In Chap. 2 I have attempted to cover the minimum system requirements for most programs. In the sections of Chap. 2, you will find discussions of many kinds of peripheral equipment available to enhance the actual modeling process for a given problem. In most cases, the right combination of equipment makes the analysis an enjoyable, rather than a tedious, task. The process of selecting a program is often a difficult choice. I have used the FEA program from ALGOR Interactive Systems, Inc. exclusively throughout this text except for one problem given in Chap. 9. Reasons for program selection are discussed in Chap. 3.

In summary, you must consider the extent of your modeling efforts, both short-term and long-term. It is best to purchase a system that will provide a comfortable environment when performing an analysis. In addition, the FEA program should fit your needs. Some programs allow purchase of modules depending on the type of analysis and depth of analysis you need. Again, one purpose of this book is to assist in your selection of the correct hardware/software combination for your particular situation.

1.5.2   The experienced user

If you are familiar with computers and the use of computers in the analysis of scientific and engineering problems, this book provides an introduction to FEA or a brief refresher to the subject. Chapter 3 may be bypassed if you regularly use FEA. If not, this chapter and Chap. 4 provide an excellent outline on the analysis technique.

1.6   Summary

The programs used in and referenced in this text have been developed and verified. The user must learn how to apply a particular program or model to the analysis problem. Note that it is not necessary to learn all of the capabilities of a particular program or model. The goal is to learn how to apply that part of a program that will, in an efficient and simple manner, solve the problems at hand.

Since Finite Element Analysis methods are widely used and have proven to be a necessary analytical tool in many areas of science and engineering, the information obtained from this book will be a stepping stone to further development of the user not just as a user, but perhaps as a program developer as well.

With many finite element programs available for the personal computer, it is difficult to select the right program. The user could review the program demos to help select the correct program. Manuals would also be helpful. However, the demo programs will feature a very professional graphics presentation and they do not let the user interface with the program to a great degree. In evaluating a program, the user must examine the following items:

1. *Program Accuracy*
   In order to validate the program accuracy, the program developers must include numerous test problems that have a classical or analytical solution. The accuracy of

each type of element and capability must be checked out, as well as the application of the program to both small and large problems.

2. *Problem Size*

Many advertisements for programs promise the solution of thousands of static and dynamic degrees of freedom. The user should verify that the program(s) will actually handle a large problem. Requests should be made of the program developers to provide the documentation.

3. *Capabilities*

Users should review their current and future requirements as to the types of problems they may encounter.

If large problems are to be solved, the program should contain solution options such as Guyan Reduction, Substructuring, or Cyclic Symmetry. Perhaps buckling capabilities or composite material analysis should be included.

4. *State-of-the-Art Techniques*

As with any technique, advances are continually being made. The user should determine if the code contains the latest in element formulation, numerical techniques, and the development of new finite element techniques.

5. *Mesh Generation*

A time-consuming task is the problem formulation and mesh generation. The program used in this book contains an excellent means of mesh generation. In order to check the mesh generation capabilities of the program that you wish to use, try generating meshes for intersecting pipes or plates with elliptical holes.

6. *Program Execution*

It is imperative that the user check the execution times of benchmark programs. Speed of execution is important, especially with large problems. Hardware considerations become extremely important.

7. *Graphics*

In order to evaluate correctly and in a timely manner, the program must be capable of displaying the undeformed and deformed structure, stress contours, and animation of the results of the dynamic analysis. A good high resolution monitor is recommended.

8. *Manuals*

In order to use the program for a long period of time, the program should come with thorough and complete documentation. This includes program use, verification problems, and a theoretical background for the program.

9. *Support*

This is a most critical item, especially for the novice user. The user must determine if the organization has the background and staff to support the program.

10. *Mainframe Capability*

For some large problems the PC-based finite element program must be capable of interfacing to some of the more well-known mainframe programs such as NASTRAN or ANSYS.

11. *Cost*

The cost of FEA programs vary and this cost does not seem to depend on the program capabilities or the efficiency of the program. Some programs have limited capability and are expensive. Other programs contain all the capabilities the user may need and the cost of these programs is quite reasonable.

1.7 What's in the Remainder of This Book?

With the above information in hand, and you as the user knowing what types of analysis you want to do and how frequently you will be performing finite element analysis, the next question is, "What type of platform will I need to perform finite element analysis?"

The remainder of the text is divided into the following chapters:

Chapter 2 covers the practical aspects of hardware requirements plus recommendations for upgrades to your existing computer systems.

Chapter 3 looks at the fundamental aspects of finite element analysis, such as setting up problems and element types.

Chapter 4 gives an introduction to the ALGOR Interactive Systems, Inc. finite element program and explains how to use the program with sample verification problems and a detailed example, both from an FEA standpoint and a conventional solution.

Chapter 5 covers the finite element methodology in somewhat more detail and presents a FORTRAN program for calculation of two-dimensional heat transfer.

Chapter 6 starts the real-world applications of FEA in a manufacturing environment with applications to tools, molds, and dies.

Chapter 7 illustrates FEA uses in the automotive industry.

Chapter 8 presents a unique view of using FEA in a musical product modification.

Chapter 9 examines finite element analysis as used in some applications in a military environment.

Chapter 10 looks at a more commercial side as it examines applications to in-home medical products.

Chapter 11 applies the finite element concept to situations within the utility industry.

The reference section contains useful references to common finite element texts and should serve as a stepping stone to further reading on the subject.

2

What Are the Minimum *Realistic* Hardware Requirements for Doing FEA on Personal Computers?

2.1 Introduction

As we move into the decade of the 1990s, the Personal Computer (PC) is and will continue to be an invaluable resource with which to perform a wide variety of computing tasks. Tasks for the PC range from word processing, spreadsheets, and report generation, to sophisticated data analysis, machine control, and computer aided design and engineering (CAD/CAE).

Convenience is the prime motivation behind the ever-increasing numbers of PCs. In general, the PC can address a wide variety of computing needs at a fraction of the cost of the much larger mainframes. While it is still true that the PC may never compete directly with the mainframe, newer and faster PCs that approach and in some cases exceed mainframe performance under special circumstances are reaching the market.

This chapter deals with the requirements needed to address FEA with the PC. The chapter begins with the most important performance item one has to consider when determining hardware requirements for performing FEA: the personal computer itself. For an existing system, I discuss the upgrade routes that I have used and the relative merits of each. Next, I discuss devices and methods with which to input and output data for subsequent analysis.

2.2   General Hardware Requirements

The basic hardware requirements for FEA are not particularly special. In order to derive the maximum benefit from a PC-based FEA system, a minimum hardware configuration is usually specified by the FEA software vendor. The rest of this chapter addresses some of the options an FEA program may require in order to perform efficiently.

2.3   The Main PC Unit

The PCs most probably found in organizations to which this text is addressed are the PC/AT, 386SX, and the 386 PC/AT bus compatibles. If you are considering purchasing newer equipment with an advanced bus technology, the following insight may be helpful.

Today the buyer has three options when selecting a computer platform on which to do FEA: (1) the classic AT bus known as the Industry Standard Architecture (ISA), (2) the newer Enhanced Industry Standard Architecture (EISA), and (3) the Micro Channel Architecture (MCA).

The first two buses trace their heritage back to the expansion bus of the original PC. The unexpected success of the IBM PC created the personal computer revolution and the need to introduce into the market machines designed to be business computers. This effort resulted in the AT with its improved 16-bit expansion bus and the ability to directly address 16 megabytes (Mb) of memory as compared to the original PC with an 8-bit bus and ability to address 1 Mb of memory.

After the AT was introduced, and as microprocessing

speed increased, the shortcomings of the AT bus were apparent. The bus speed was too slow to keep up with the memory needs of 80286 microprocessors running faster than 8 MHz. Some compatible computer makers separated the microprocessor memory from the general expansion bus. Through a direct route to the microprocessor, memory could operate at speeds as fast as the memory chips would allow. All current high performance PCs running on the ISA bus use this theme or a variation of this theme.

IBM's solution to the AT bus was the Micro Channel Architecture. This bus design included greater data-moving capability and other innovations. This architecture is incompatible with PC and AT expansion boards. The bus may offer technological advances, but the designing of expansion boards and compatible machines is difficult, to say the least, according to industry sources that I have contacted. In addition, the royalties that must be paid to IBM for using the MCA proprietary technology does not sit well with other manufacturers.

EISA was developed in response to the MCA channel by a consortium of leading IBM-compatible manufacturers. The purpose was to provide the power of the MCA bus without the disadvantages. EISA's features essentially duplicate those of the MCA. For board designers, perhaps the best aspect of the EISA bus is the backward compatibility with the ISA bus. Even though the EISA board designs are as complex as the MCA board designs, ordinary board designs still work for ordinary purposes. Engineers can get "simple" interface boards to market without much difficulty.

The most controversy between the two advanced bus designs lies in "improved" performance speed. This could be the very reason to spend the extra money to buy one of these systems. However, specifications and reality often do not coincide. While pure performance numbers specified by the manufacturer are usually true, these performance numbers may not be relevant to your everyday use of the system. Testing by major test organizations has shown that for single-user DOS applications the ISA bus is still the best alternative from a price/performance perspective. Tests of all three buses have shown that information is moved across each bus at about the same speed. However, for shared resources, such as in a Local Area Network (LAN) environment, the MCA has a slight edge on the EISA bus in performance, and a great advantage over the ISA bus. As more peripherals become available for the advanced architecture, the performance increases over the ISA bus may contribute to the user selecting one bus over the other.

For the majority of readers, this section will provide upgrade directions based on the premise that the user has an 8086, 8088, or 80286 machine and wishes to move into the 80286, 80386, or 80486 processing environment.

The problems in this book were solved on an 80386 computer clocked at 25 MHz with a cache of 64K. For an all-around, reasonably priced alternative, this is the machine that I would recommend for FEA calculations, although more horsepower is better for any type of intensive numerical calculations. However, there are many alternatives for achieving optimal performance from your existing computer platform if economics is a concern. The following sections describe these upgrades.

2.3.1   8086/8088 performance upgrades

Can you run FEA on a PC/XT or compatible? Of course the answer is yes. As a minimum you will need the following in your system:

1. Basic XT machine

2. 8087 math coprocessor

3. 640K of RAM

4. 10 Mb or larger hard disk

5. CGA for color graphics (EGA minimum preferred)

It is apparent that, depending on the FEA program size and problem storage requirements, the model size will be limited to some extent. However, this is a very workable system and can provide insight into your problems if you are working on problems that are not relatively complex. For a 10 Mb system, you may be limited to approximately 200–250 three-dimensional plate elements.

There are options to replace the 8088/8086 processor with items known as 80286 Accelerator Boards and 80386 Accelerator Boards. These boards fit into one of the 8-bit or 16-bit slots on the motherboard with a cable to connect the board to the existing 8088/8086 chip socket. For some CPU-intensive calculations, the system speed can be increased up to 600 percent. However, for most applications that require numerous hard disk read/write actions performance will be on the order of 130 to 180 percent. A 80386 Accelerator Board should perform from 200 to 300 percent faster. The improvement is still worth the investment from a time viewpoint.

The reader should review some of the more popular computer magazines and check discount computer stores for availability and the best price of various accelerator boards. Also note that for computationally intensive applications, a numeric coprocessor will be required and will be an additional expense. The reader may wish to use the Intel series of math coprocessors or he or she may want to investigate the newer high performance math coprocessors, such as the IIT 80×87, USLI, or perhaps the Cyrix line of math coprocessors.

The other major upgrade to an XT system is the replacement of the system motherboard with an 80286 or 80386 motherboard. This option may be slightly more expensive than the accelerator board option, depending upon where you purchase the motherboard; however, this option does give the advantage of providing increased system throughput due to the increased bus width from 8-bit to 16-bit for the ISA bus. Comparative shopping is the key to obtaining the best performance-to-price ratio. Table 2.1 lists some available accelerator boards.

Any system can be improved upon from a performance standpoint by upgrading peripherals such as fast access hard drives, additional memory for caching, and high performance controllers. These options do not give as much improvement for the dollar as the above-mentioned items, but, when coupled with a replacement motherboard, you will wonder how you did without them for such a long time.

2.3.2   80286 performance upgrades

If you are currently using an 80286 computer, you may still wish to upgrade your computing platform. Certainly there is the system peripherals upgrade that improves performance. But again the most performance improvement short of replacing the entire machine is to add an accelerator board or replace the motherboard.

There are several manufacturers of accelerator boards. For this text, I chose to test an 80386 Accelerator Board from MicroWay of Kingston, MA. This board is a full slot width, add-in board that replaces the 80286 in an IBM AT or compatible with its on-board 80386. The board includes sockets for one, two, four, or eight megabytes of 32-bit memory, an Intel 80387, Cyrix FasMath 83D87, or Weitek 3167 numeric coprocessor, and 64K or 256K of high speed cache memory. The Number Smasher-386 functions as an asynchronous emulator with shadow memory. It is compatible with any zero- or one-wait-state 80286 system having clock speeds from 6 MHz to 12 MHz.

TABLE 2.1   Accelerator Boards and CPU Ratings

Intel Above Board 286+	8 MHz
Microsoft MACH 20	8 MHz
Orchid Tiny Turbo 286	12 MHz
Quadram QUAD386XT	16 MHz
SOTA 286i	12 MHz
Tallgrass Tech. Short Cut	12.5 MHz
MicroWay FASTCache-SX	16 MHz

Using the Whetstone benchmark as a basis for comparison, the Number Smasher-386 runs approximately 65 times faster than a 6 MHz AT. All 16-bit 80286-based applications and operating systems run with this board. In addition, 32-bit protected mode code written for the 80386 as well as other specific 80386-based operating environments runs without problem.

The bottom-line performance data suggests that the Number Smasher-386 provides 80286 users with the ability to take advantage of the speed, power, and performance of the 80386 while maintaining full compatibility with existing AT hardware and software.

While the add-in board does provide a substantial amount of power for the user, the cost of the board is such that you may wish to replace the motherboard in your 8088/8086 or 80286 computer. For this book and my personal use, I replaced my 80286-10 motherboard with an 80386-25 motherboard with 64K Cache and 8 Mb RAM. This combination was less costly than using an accelerator board. The performance of the replacement motherboard is as good as, or better than, the plug-in accelerator board. However, to optimize a system, high performance peripherals have to be included. I can make a "slow" 80386/486 system by loading the system with low performance peripherals. There are many combinations of peripherals for a given system which will provide the performance that you want.

If you are considering upgrading an 80286 system to a faster 80286 system, I would not recommend it. Cost-wise, an 80386-SX system would be the choice if the 80386 is not an option. If you have a 16–20 MHz 80286 system equipped with reasonably performing peripherals, you will probably not want to spend the extra money to upgrade the system. The user has to look at percentage uses for analysis versus other tasks done on the machine.

You should also consider that newer versions of programs are being written to take advantage of the 80386/486 architecture.

2.4   Input Devices

Our discussions of the various input and output devices will address mostly those aspects which pertain to the input and output of graphical data. Textual and numeric data would generally be input or output for the purpose of transferring data to an auxiliary program. As an example, one may wish to send a numeric data file to a mainframe computer after using a PC to build the model. The major thrust behind using any input device is to provide a fast way of entering data. To this end, it is clear that those devices which provide a graphical (mouse, light pen, or digitizer) rather than alphanumeric (keyboard) input are more desirable.

2.4.1   Keyboards

From the standpoint of graphical data input, keyboards are probably the most cumbersome method of entering graphical data. This fact is obvious, since anyone who has ever typed reams of data on a keypunch machine or at a CRT knows how tiresome this is. Still, the keyboard as a method of graphical data entry has a place even with the existence of simpler methods. Sometimes, a small change to a graphics data file may be easier via keyboard entry rather than executing a special program to perform the same task. One of the keyboard's drawbacks is the speed with which one may enter data. Data entry through a keyboard is the least costly method since it is an integral part of the PC. Keyboard data input is primarily used to issue commands and to enter a few system variables into the FEA program.

2.4.2   Mice

If you already have a mouse or are familiar with how a mouse is used by a computer system, you may wish to skip this section.

The entry of graphical data via a mouse is a definite step in the right direction. The mouse adds the convenience of speed and ease of use as well as being low in cost. There are generally two types of mice available.

The first method uses what is called the inverted track-ball. As you move the mouse, the ball rotates, moving rollers inside the mouse case. The relative movement of the cursor is in the horizontal and vertical direction on the screen. Because mechanical mice are analog devices, this method has the advantage of being able to operate on almost any surface. The drawback to using a mechanical mouse is that it easily picks up dirt from the surface on which it moves, and should be cleaned regularly.

The second method uses an optical pad. With this method, two pieces are needed: the mouse and a tablet which has been etched with a grid pattern on its surface. This technique of implementing a mouse is most useful when a limited amount of area is available, since the mouse works only on the grid surface and this is usually about 6 inches on a side in size. There are two openings on the bottom of an optical mouse. Light emitted from a diode inside the mouse shines out through one of these openings, reflects off the special pad, and is read through the other opening. As the mouse is moved, the reflected light is interrupted by the grid marking on the special pad. The grid lines in the X and Y directions on the pad each reflect only a certain phase of the light emitted by the mouse. The number of times the light is interrupted in each polarized direction is translated to the amount of movement of the cursor in the horizontal and vertical directions on the screen.

This is the most cost-effective method of imputing FEA models and manipulating the models in PC-based programs.

2.4.3   Tablets

Tablets are probably as efficient a method of data entry as using a mouse. A tablet consists of a special surface and a locating device, and provides a surface of a specific size which is mapped to the screen of the CRT. Selecting a point on the tablet surface corresponds to selecting a specific point on the screen. The locating device is either a hand-held cursor (sometimes called a *puck*) or a pencil-like *stylus*. Most digitizer tablets today operate using ultrasonics. The stylus is placed at any arbitrary location on the tablet; this position is then determined through the use of two micro-phones. The implementation of three-dimensional digi-tizers is a simple extension of this principle. By adding another microphone in the Z direction, we have accom-plished our task. The spatial resolution of the digitizer is typically 0.001 in. The digitizing surface area ranges from 6 by 6 inches to 4 by 6 feet.

2.5   Data Output Devices

Data output devices are used to present or transmit (to another program) the results of an FEA analysis. As with input devices we are, for the most part, interested in graph-ical output devices. The following section will describe var-ious graphic display methods and attempt to provide the reader with the basic advantages and disadvantages of each method.

Before we begin, two quantities used to describe graphi-cal output devices should be addressed. These are Spatial resolution and Spectral resolution. Spatial resolution addresses the "fineness" of detail which we perceive when we view an object. Spectral resolution addresses the infor-mation content of a discrete unit (usually called a Pixel) of information contained within the object of interest.

2.5.1   Graphics boards

The graphics display board within the PC is used to gener-ate a video signal, either analog or RGB, for display on the PC's system monitor. One of the most important and often overlooked aspects of building an FEA model is the resolu-tion of the graphics display board. A standard Color Graphics Adapter (CGA) provides reasonable graphics capability, but at a 320 by 200 pixel resolution a user will find the displayed image somewhat lacking in quality. The displayed image of the CGA will produce jagged edges for diagonal lines and limit the size of the displayed model. To overcome this, manufacturers of graphics boards have produced the Enhanced Color Graphics (EGA) board. This board provides up to 640 by 350 pixel resolution. The standard now is the VGA graphics boards providing even more resolution (up to 1280 by 1024) to further enhance displayed graphics. One of the advantages of higher resolution is the ability to draw larger models, since more picture elements are available within the displayed area of the CRT. Another feature of many of the add-in graphics boards is the ability to display more than 16 colors (or gray shades) up to 256 colors simultaneously. The number of colors or gray shades is not as important when defining a model because only one intensity is needed to draw the model structure, but the number of colors (or gray shades) does add to the quality of the video output and can be important when viewing stress, temperature, velocity, or pressure output files. While generally not a problem, the refresh rate of the graphics board should be considered when selecting a display board. Refresh rate is the rate at which the graphics board updates the video information contained in video RAM to the CRT display. In general, the faster the refresh rate the more solid the image appears flicker-free. A refresh rate of 30 Hz will in most cases appear to flicker while a 60 Hz or higher refresh rate is perceived as a steady image.

Many FEA software vendors supply driver programs which support many of the popular high resolution graphics boards.

2.5.2   CRT displays

Depending upon your personal requirements, the CRT can be the least costly or most expensive part of the FEA/PC system. The resolution of the CRT display can cover a wide

range. Many vendors supply CRTs from 12 to 27 inches
(diagonal measure) for high resolution display (1280 by 1024
pixels). CRT displays have other performance criteria which
provide a measure of the viewing quality one perceives. The
persistence of the phosphor used in the manufacture of a
CRT determines the decay rate of the image displayed. If the
phosphor's persistence is too short, fading of the image will
occur and, depending on the refresh rate, flicker may
appear. Flicker is an undesirable attribute and is reduced in
one of two ways: (1) increase the refresh rate, and (2)
increase the persistence of the phosphor in the CRT.

2.5.3 Printers

The printer may be used to provide alphanumeric data as
well as graphical data. Although a printer's spatial resolution is limited, it does provide an inexpensive method of
producing hard copy results of an FEA analysis. Printer
resolutions range from 60 dots per inch for low end 9-pin
dot matrix printers to 300+ dots per inch for LASER printers. Several newer printers also offer multicolor capability
at very reasonable prices. I would recommend at least
investing in a color dot matrix or color inkjet printer for
impact in presentation of analysis results.

2.5.4 Plotters

The plotter has been a desirable method for the display of
graphical results. A plotter can range in cost from several
hundred dollars for "A" size to several thousand dollars for
plots from "A" size to "E" size or larger. A plotter offers
greater resolution than a printer (generally a function of
cost) and speed (also a function of cost).

For large drawings, plotters are what you need to use.
For the types of output expected in the manufacturing environment, a good LASER printer will produce equivalent or
better results with the appropriate printer drivers.

2.5.5 Software

For presenting the results of your analyses, some form of screen capture software is required for making hard copy (with your printer or plotter), overhead transparencies, or slides. A program that I recommend is Pizazz Plus by Application Techniques, Inc. (10 Lomar Park Drive, Pepperell, MA 01463). With this program and with your given computer and printer combination, you will be able to capture the images that you see on the screen to hard copy. This can be done in color or black and white. Another method that I recommend is taking slides directly off the screen image. I have found that excellent presentation-quality slides can be produced using a 35mm camera with a telephoto lens to crop the screen image, and Kodak Ektachrome 200 or equivalent.

2.6 Performance

2.6.1 Clock speed

CPU clock speed is a measure of performance when comparing like bus structures. If we limit our discussion to the PC, we may make this comparison. If we discuss other platforms, such as the Macintosh by Apple or the SUN SPARC Station, then we must make comparisons based on overall system throughput.

The bottom line is to have the fastest clock speed CPU possible.

2.6.2 Add-in coprocessors

Several manufacturers of PC-compatible products provide add-in hardware which provides capabilities approaching mainframe performance. These hardware add-ins are typically called array processors and for the most part do indeed function in the same manner as their bigger brothers. The drawback to these devices is that they are generally not supported by the FEA models currently on the market. Therefore, adaptation to one of these models may or may not be possible.

2.7 System Enhancements

2.7.1 The 80×87 math coprocessor

The 80×87 math coprocessor chip is typically specified as a required piece of hardware by most FEA software vendors. This is due to the mathematical intensity required of FEA models. The coprocessor chip can increase the speed of some calculations by two or three orders of magnitude.

The new 80486 CPU from Intel has the coprocessor built in. Often the cost of a 33 MHz 80386 with an 80387 coprocessor is not much less cost-wise than a 25 MHz 80486. Your throughput for the 80486 will be greater.

2.7.2 Software additions

Hardware determines the baseline performance of a PC system but some software packages offer additional convenience. CAD software, such as AutoCAD, VersaCAD, or CADKEY can provide a faster method of model data entry than most FEA programs. One simply draws the model using a mouse or digitizer, then submits the resulting file to a conversion program generally provided by the FEA software vendor.

2.7.3 Hard drives and hard drive controllers

Among the most important parts of a system are the hard drive and the hard drive controller unit. For PC-based applications, due to memory limitations inherent in the DOS structure, it is necessary to have the largest and fastest hard drive/controller combination that you can afford. In addition, the proper controller for the hard drive that provides from 800,000 bps to 10 million bps is necessary since the hard drives are used as virtual memory devices and access to the hard drive is often quite frequent. The larger the problems, the more frequently the hard drive is accessed. If the FEA program is optimized for the 80386 or 80486 machines and has additional memory beyond 2 Mb on board, the number of times for hard drive access will be reduced. There are a number of manufacturers, such as Conner, Maxtor, Micropolis, and Seagate, that provide large hard drive/high performance units. I recommend at least a 100–150 Mb hard drive with average access times of less than 20 ms. Hard drives of this caliber are not inexpensive. Units range in price from $500 to $1200. The time you will save for heavy analysis will pay back the cost of the drive in a reasonable amount of time.

If you are hesitant about adding or replacing a hard drive, there is an alternative that is quite simple and that was used for evaluation during a portion of the development of this text. Plus Development Corporation (1778 McCarthy Blvd., Milpitas, CA 95035-7421) offers a true plug-and-play unit in its Hardcard II XL series. I used the 105 Mb version. This device is a complete hard drive system. All controllers and drive electronics are integrated on a single PC/AT card. With an installed 64K disk cache, the effective disk access speed hovers around 9 ms for frequently used data. Otherwise, access times are in the high 15 ms to low 16 ms range. If you have another drive, the unit installs in a standard 16-bit slot inside the PC and automatically configures itself as the next available drive.

The drive is a high-performance add-in peripheral. Installation is quite straightforward with ample instructions for many configurations. The user has the ability to change address ports depending upon the user's current configuration. It is best to read the manual and especially the readme file on the installation diskette. For example, the information in the "readme" file includes tips on installing in computers not listed or discussed in the manual and how to use the Hardcard II XL with memory managers such as QEMM from Quarterdeck. The Hardcard II

XL is compatible with all 286/386/486 PC systems and with the following operating systems:

DOS 3.x, 4.x, 5.0

Novell Netware 286 Advance and SFT

3Com3+, 3Com3+Open

Microsoft Windows 3.0, OS/2

SCO XENIX 286 System V

SCO UNIX 386 System V

The Hardcard II XL behaved as a normal drive. The utilities used (PCTools V7.1 from Central Point Software) recognized the drive as drive D. All utilities worked without any problem on the drive.

All finite element software was loaded onto this drive. It was quite evident that the quickness of the drive reduced overall analysis times by at least 30 percent when frequent hard drive access was required for the program. All graphics for the models appeared twice as quickly as from the original drive used in the system.

For this type of analysis, the user needs a fast hard drive. This is certainly what you get with the Hardcard II XL. I have no hesitation in recommending this unit as a first line upgrade for any 286/386/486 system, either as a primary drive unit or as a secondary drive. For users needing to upgrade systems for such analysis as is herein discussed, no better solution currently exists for the price/performance ratio.

The Hardcard II XL comes with a two-year, end-user limited warranty. From a review of previous data concerning reliability, there seems to be a very high level of customer satisfaction, with failures from infant mortality or other problems being very small. In addition, the price is well in line with the competitive capacity hard drives and requires a serious look from those contemplating a replacement or additional hard drive for their system.

2.7.4   Front-end CAD for FEA applications

Most FEA programs today generally include a preprocessor that is a basic three-dimensional CAD system. Although not a full-featured CAD for dimensioning, these preprocessors have the ability to create accurate three-dimensional models of the object set for analysis. This allows the user to rapidly build a model without using the keyboard method of entry of building a model. I have worked in the past with programs where model building was required node-by-node by imputing xyz coordinates for each point. To see how the model was progressing, I would have to exit the input program and then run a graphics viewing program. This was a very tedious and iterative process. Fortunately, today's programs have much better input capabilities for three-dimensional model input. This feature allows for instantaneous feedback in the modeling process. The actual building of the model is simpler and even fun.

If you have an existing CAD package such as AutoCAD 10 or 11, for example, you may wish to design on that software and translate the DXF file to the FEA package. All major FEA programs have translators for the various CAD packages. Using this feature and the ability to translate back to the existing CAD package that you are using provides an ideal setting for an iterative design process between the designer, engineer, and manufacturing engineer. This allows design problems to be detected early on and corrected before committing to prototype or production.

The FEA preprocessor can be used as a "concept development" station before releasing your CAD sketches for detailing. All preliminary design and modeling can be done and changes made for a shorter route to production.

Other than cost and possible duplication of features, the CAD tool supplied by the FEA programs will complement your existing design efforts and in essence give you another design station in your organization.

3

Fundamentals of Applied Finite Element Analysis

3.1 Introduction

The finite element method has been used for many years to solve complex problems in many fields, including aerospace, automotive, civil, and mechanical engineering. Early in the development and implementation of finite element codes, basic linear static and dynamic problems had to be solved on large mainframe computers. Today, very sophisticated and intricate problems can be solved on smaller computers such as the personal computer. This fact puts finite element analysis capabilities at the disposal of engineers in many disciplines.

3.2 Basic Concepts

The Finite Element Method is an analytical tool for performing stress and vibration analysis (both linear and nonlinear), thermal (steady state and transient), and fluid flow (laminar and turbulent) analysis of systems and structures. A typical analysis using the finite element technique requires the following information:

1. Nodal point spatial locations (geometry)

2. Elements connecting the nodal points

3. Mass properties

4. Boundary conditions or restraints

5. Loading or forcing function details

6. Analysis options

This information only directs the user toward the final solution for the particular problem in question. The user must, in an intelligent manner, interpret the results and apply the results to the actual system behavior. This is a goal of the application of finite element analysis to practical scientific and engineering problems.

Certainly there are other goals beyond the system behavior. If the user is addressing, for example, a structure of some nature, one goal could be to develop an understanding of the structural integrity of the system. If an optimum design of a structure is wanted, then the user can use the finite element method to examine how a structure (system) responds to design modifications. Furthermore, from a cost standpoint, the use of finite element modeling to assess or simulate system testing or test results presents a goal of product development cost reduction.

Can finite element analysis really be done on personal computers? In reviewing the literature, there appear to be three opinions concerning the use of personal computers and finite element analysis as a unit.

The first opinion views the combination as merely a tool with which to learn basic finite element techniques and to solve "small" problems in order to develop a qualitative insight to a problem.

The second opinion is that full-featured linear finite element analysis is possible on personal computers.

Finally, the third opinion is that a full-featured finite element analysis (linear and nonlinear) is possible on enhanced versions of PC/ATs and compatibles. I agree with this opinion and, after using a PC-based finite element program, I believe that you will agree also.

The following sections examine in detail these aspects of finite element modeling and the details in the various components of a finite element model.

3.2.1 The physical problem

To start a finite element analysis, one must first survey the item or situation to be modeled and all boundary conditions and restraints. A careful problem review may lead to an application of the precise element; otherwise, one may be tempted to use too complex an element for the problem under consideration. The use of the inappropriate element can lead to long computational times and solutions that are not accurate. Furthermore, the answer, although possibly correct, may have been arrived at with fewer and less complicated elements.

The use of FEA relies on the user's understanding of the underlying theory of the types of elements and their applications, and on the experience of the user in the field in which he or she is working. It is suggested that the users of this text may wish to build small verification models particular to their field of expertise. This type of building and learning affords the user the confidence level to pursue more complicated models.

3.2.2 Criteria for using FEA

When is it necessary to use FEA, and when should one use some other method? A new initiate to FEA may wish to solve every problem that may arise using this technique. The more one uses and explores the capabilities of an FEA program, the better able he or she is to adapt the techniques to other problems. However, one must not lose sight of the fundamentals behind the analysis of a problem using FEA. The solving of simpler problems, on occasion, by conventional methods, is encouraged.

One should determine if FEA is the quickest and most cost-effective solution to product design and development. If this is not the case, then FEA should not be done.

Inexpensive hardware prototyping and the availability of reliable testing generally precludes the use of FEA.

FEA can be used as the necessary tool for the design and development when any of the following conditions (not an exhaustive list) apply:

1. The product is not a fabricated product.

2. Tool modifications may be required.

3. The production item would not perform as a fabricated part.

4. The design process calls for material optimization.

5. Prohibitive model construction costs are involved.

6. Prohibitive (expensive and/or dangerous) test requirements are anticipated.

7. Legal aspects of product design are important.

It is very important to use FEA or some other tool such as prototyping as early in the design stage of a project as possible. An insight into possible problems will provide a more reliable product.

3.2.3   Using the correct engineering units

FEA programs do not contain a fixed set of units within which to work. The user selects the units that are best suited to the program. The proper selection of units and adhering to the selected set of units are important in achieving the correct solution to a problem. Table 3.1 lists the various quantities and the units particular to the quantities in terms of mass (M), length (L), and time (T). Common English and metric units are given.

3.3   Discretization

Discretization of the physical problem into subdivisions or regions is the first step in any analysis. If a structure or system is discretized, the meaning is that the system is

TABLE 3.1   English and Metric Units

Engineering Units for Commonly Encountered Variables

Length	L	inch	meter
Mass	M	lb·sec^2/inch	kilogram
Time	T	second	second
Area	L^2	inch^2	m^2
Volume	L^3	inch^3	m^3
Velocity	LT^{-1}	inch/sec	m/sec
Acceleration	LT^{-2}	inch/sec^2	m/sec^2
Rotation	—	radian	radian
Rotational Velocity	T^{-1}	rad/sec	rad/sec
Rotational Acceleration	T^{-2}	rad/sec^2	rad/sec^2
Frequency	T^{-1}	Hertz	Hertz
Force	MLT^{-2}	pound	newton
Weight	MLT^{-2}	pound	newton
Mass Density	ML^{-3}	lb·sec^2/inch^4	kg/m^3
Young's Modulus	ML^{-1}T^{-2}	lb/inch^2	N/m^2
Poisson's Ratio	—	—	—
Shear Modulus	ML^{-1}T^{-2}	lb/inch^2	N/m^2
Moment	ML^{2}T^{-2}	inch·lb	N·m
Area Moment of Inertia	L^4	inch^4	m^4
Torsional Moment of Inertia	L^4	inch^4/rad	m^4/rad
Mass Moment of Inertia	ML^2	inch·lb·sec^2	kg·m^2
Stress	ML^{-1}T^{-2}	lb/inch^2	N/m^2
Strain	—	—	—

represented by a discrete grid or node points connected by elements. The finite element method is a discrete representation of a continuous physical system. It is believed that a theoretical basis to perform the subdividing of a region into a set pattern of elements does not exist. Engineering judgment is the key at this stage of the analysis. Decisions must be made regarding the number and the size and shape of the subdivisions. This solution process is balanced against the selection of the proper elements that produce accurate and useful results in a reasonable amount of computational time and computational efficiency.

The common rule of thumb in discussions about discretization is: the greater number of nodes and elements, the more accurate the solution. In theory, this assumption is correct. I would suggest using as few nodes as possible that follow the purpose of the analysis and are consistent with the element chosen to model the situation. If a detailed stress analysis is a requirement, then the nodal density in the regions of large stress gradients is by requirement increased. For example, increased mesh densities are required in regions of applied forces and in regions surrounding cutouts or sharp corners. If only deflections are required, then perhaps even fewer node points are required.

3.3.1 Node points

The finite element method discretizes a structure by defining node points connected by elements. The nodal points are located in space relative to a coordinate system often referred to as the *global coordinate system*. Proper choice of the origin when developing any model can often simplify model definition and input data. However, with a graphical preprocessor, origin selection is not as critical as it once was for some problems.

Each node (and each element connected to the node) is initially free to move about in space. The motion is represented by three translations (assuming a cartesian system) in the x, y, and z directions and by three rotations (θ_x, θ_y, and θ_z) about the point centroid. Each direction of motion is termed a *degree of freedom*. For each unconstrained node there are six degrees of freedom. As the model is developed, some nodes will be restricted in their motion. These restrictions or boundary conditions will represent actual physical restraints on the system.

3.3.2 Coordinate systems

Node point locations are specified relative to a particular coordinate system. The global coordinate system is usually the cartesian coordinate system with axes labeled X, Y, and Z. The three axes are mutually perpendicular. Note that all coordinate axes, nodal displacements, and applied forces adhere to the "right-hand rule" of orientation.

3.4   Selecting the Proper Elements

For the wide variety of problems that one may face, there is no one element that can satisfy all requirements. There are specific elements for one-dimensional, two-dimensional, and three-dimensional problems. There are elements for pure stress, stress plus bending, thermal elements, fluid elements, etc. There are triangular plate elements, quadrilateral plate elements, eight node solid brick elements, etc. The purpose of this section is not to provide an exhaustive look at the different types of elements, but to look at the more common elements generally used in solving typical problems.

A finite element analysis is better developed if the user has a concept of the expected behavior of the system being modeled. This expected behavior of the actual system comes from prior knowledge of the system or similar systems and from calculations performed on simplified models. There are no exact formulas or rules to aid in this development process and/or selection of elements.

3.4.1   Element types

The following section describes the more commonly used elements for structural, thermal, and fluid analyses. Individual programs offer these and other advanced elements. The user should determine his or her requirements and needs accordingly.

3.4.2   Commonly used structural elements

3.4.2.1   Three-dimensional two node truss. The three-dimensional two node truss element is perhaps one of the simplest elements available. This element has a cross-sectional area and is shown in the following Fig. 3.1 as a line segment. There are two nodes, one at each end. Each node has three degrees of freedom and is not capable of carrying bending loads. Loads applied to this member include nodal

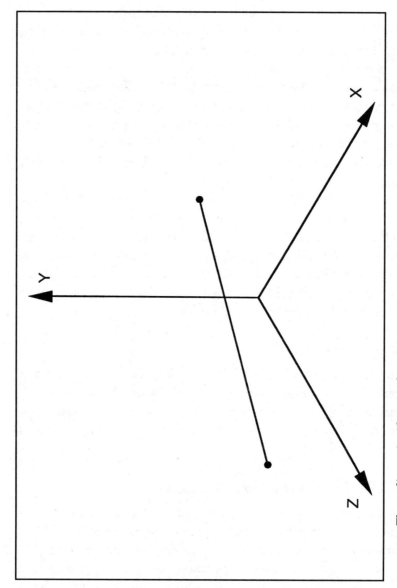

Figure 3.1 Three-dimensional truss element.

forces, thermal, and uniform acceleration in any or all directions. The stress is assumed to be constant over the entire element. This element is used in problems involving two force (truss) members such as in the modeling of towers, bridges, and buildings.

3.4.2.2 Three-dimensional two node beam. The three-dimensional beam element, unlike the two node truss element, allows bending. Typical loading can consist of moments and forces at nodes, moments and forces at intermediate locations, thermal, continuous and intermediate distributed loading, fixed-end forces, and accelerations in any or all directions. The user must specify the Moments of Inertia in both the local x and y directions as well as the shear areas in the local directions. The degrees of freedom are translation in the local x, y, and z directions as well as rotations in each of the local directions. The beam must not have a zero length or area. The moments of inertia, however, may be zero. The beam can have any cross-sectional area for which the moments of inertia can be computed. This element is used where a typical beam may be used in real-world problems. For example, this element should be used in the analysis of three-dimensional frames such as bridges and powerline towers. This element is illustrated in Fig. 3.2.

3.4.2.3 Three-dimensional four node membrane plate. The three-dimensional four node plate (and variations of this element) are very useful in the modeling of structures where bending (out-of-plane) and/or membrane (in-plane) stresses play equally important roles in the behavior of that particular structure. A careful analysis of the structure and the reaction of the structure to the applied loads should be given much consideration.

Each node has six degrees of freedom, translation in each of the local axes, and rotation in each of the local axes. The user must supply the correct plate thickness. The reader

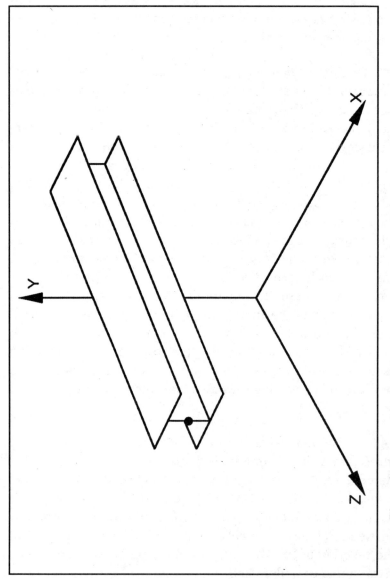

Figure 3.2 Three-dimensional beam element.

should refer to the appropriate sections of this book that are concerned with solution accuracy. Membrane elements are used in the design of fabric-like structures such as tents, and roofs of athletic structures such as stadium domes. This element is illustrated in Fig. 3.3.

3.4.2.4 Two-dimensional four and three node solid.

The two-dimensional four node solid is also referred to as the isoparametric quadrilateral element. This is perhaps one of the more common elements that are used in two-dimensional stress problems and natural frequency analysis problems for solid structures. The element is assumed thin to the degree that the stress magnitude in the third direction is considered constant over the element thickness. In addition to being used as a biaxial plane element for stress, the element can also be used as a plain strain element. There are two translational degrees of freedom and no rotational degrees of freedom. Temperature dependent, orthotropic material properties can be defined. Figure 3.4 shows this element.

3.4.2.5 Three-dimensional eight node solid brick.

The three-dimensional eight node solid brick element has three translational degrees of freedom per node. No rotational degrees of freedom are allowed. This element, shown in Fig. 3.5, is most useful in the modeling of items such as bolts, washers, and heavy metal casings, or practically any solid structure. This is a linear element in that the gradients are not constant but are a linear function of one of the coordinate directions. Isotropic material properties and incompatible displacement nodes are used in the formulation. These elements are used in the modeling of items such as wheels, turbine blades, and flanges.

3.4.2.6 Three-dimensional plate/shell element.

These are three and four node elements formulated in three-dimensional space with five degrees of freedom defined for each

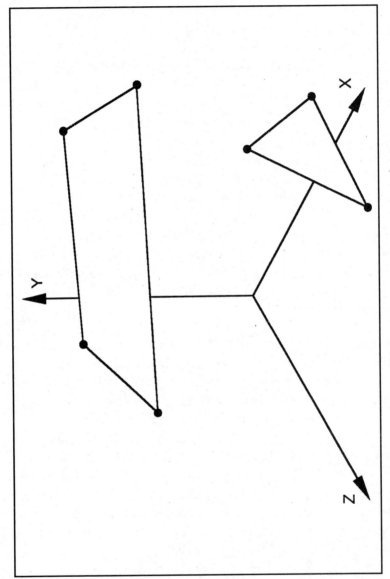

Figure 3.3 Membrane three-dimensional plane stress.

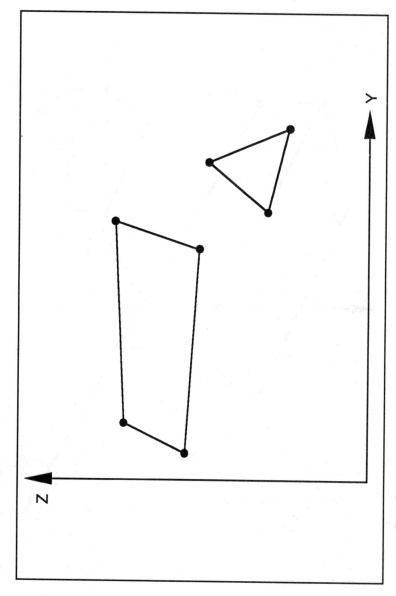

Figure 3.4 Two-dimensional solid elasticity.

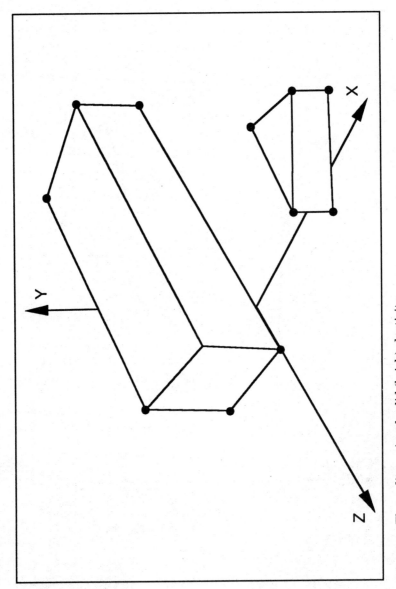

Figure 3.5 Three-dimensional solid (brick) elasticity.

node. There are three translations and two rotations which produce out-of-plane bending. In-plane rotation is not defined for plate elements. An optional in-plane rotational stiffness is automatically added to the nodes of each element. Orthotropic material properties can be defined. These elements are used in the modeling of items such as pressure vessels, electronic enclosures, and automotive body parts. Figure 3.6 illustrates this element.

3.4.2.7 Three-dimensional boundary element. The three-dimensional two node boundary element is allowed translation in each of the local directions. As shown in Fig. 3.7, the only quantity that must be known is the stiffness of the element. This element can be quite useful in cases where there is no clear-cut "fixity" of a particular node. Again, as with other elements, some trial and error may be required and a substantial amount of engineering judgment may have to be brought into play. Used in conjunction with all other elements, boundary elements rigidly or elastically support a model and enable extraction of support reactions. At either node, or both nodes, a specified rotation or translation can be imposed.

3.4.2.8 Commonly used fluid/thermal elements

3.4.2.8.1 Two-dimensional elements. Figure 3.8 illustrates the two-dimensional Isoparametric Thermal Solid. This element can be used as a biaxial plane element or as an axisymmetric ring element with a two-dimensional thermal conduction capability. The element has four nodes. Each node has a single degree of freedom (temperature). The element is applicable to two-dimensional steady state and transient analyses.

If the model containing the isoparametric temperature element is to be analyzed structurally, the element should be replaced by an equivalent structural element.

The element must not have a negative or zero area. As with all quadrilateral elements, the node numbering must be counterclockwise relative to the local coordinate system.

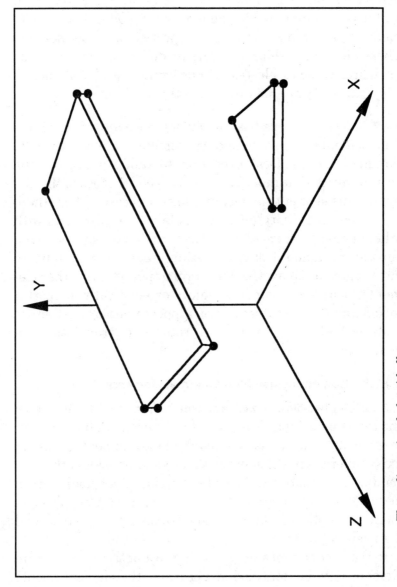

Figure 3.6 Three-dimensional plate/shell.

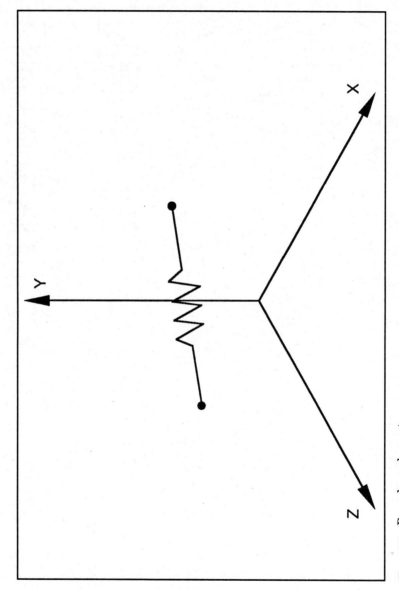

Figure 3.7 Boundary elements.

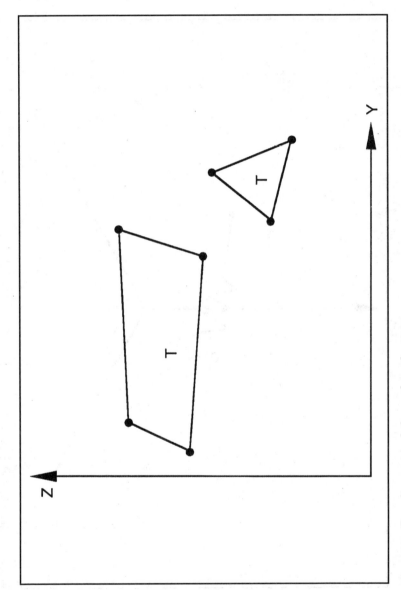

Figure 3.8 Two-dimensional thermal conductivity.

Examples of designs modeled with these elements are pressure vessels, printed circuit boards, and cooling fins.

3.4.2.8.2 Three-dimensional thermal elements. These elements are formulated in three-dimensional space. Material property data may describe orthotropic, temperature-dependent behavior. Loading can be through internal heat generation, surface convection, radiation, and constant nodal temperatures. This element is shown in Fig. 3.9.

3.4.2.8.3 Temperature boundary elements. These one node elements, shown in Fig. 3.10, provide a method of specifying a temperature boundary condition on a node. The attachment node defines the node which will receive the boundary condition. This element is used in conjunction with the two-dimensional or three-dimensional thermal elements.

3.4.2.8.4 Two-dimensional fluid element. The two-dimensional fluid element is a modification of the isoparametric solid element. This fluid element is well suited for modeling hydrostatic pressures and fluid/solid interactions. The element is defined by four nodal points having two degrees of freedom in the local x and y directions. The area of the element must be positive and the numbering of the nodes must be counterclockwise in the coordinate system. The amount of flow permitted is limited to that which will not cause significant distortions in the element. When used for a static application, the free surface must be input as flat.

3.5 Elements, Nodes, and Degrees of Freedom

3.5.1 General discussion

Many commercially available programs advertise and emphasize the number of nodes available to the user for modeling purposes. While this information is necessary, the feature that the user must focus on is a combination of the

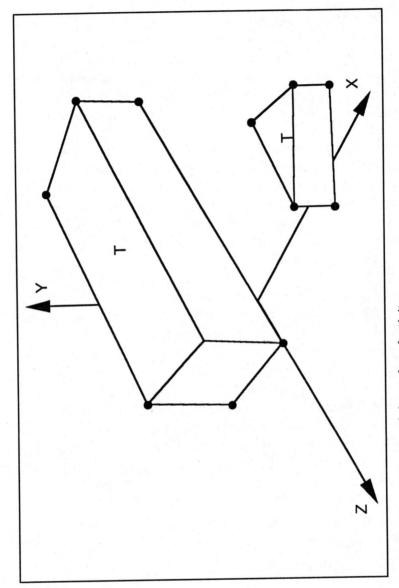

Figure 3.9 Three-dimensional thermal conductivity.

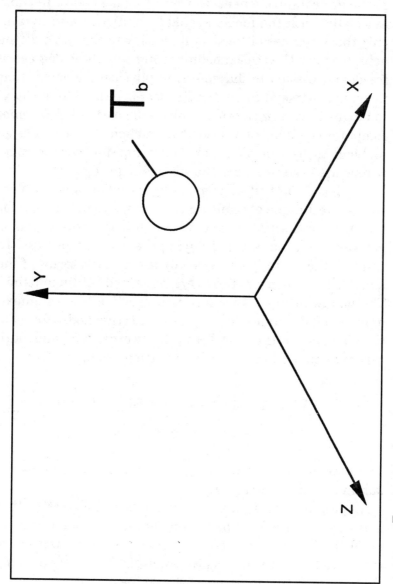

Figure 3.10 Temperature boundary element.

number of nodes, the type of element chosen for the analysis, and the degrees of freedom associated with a node of a particular chosen element. However, the degrees of freedom associated with the model should be sufficient and contain only those degrees of freedom necessary to characterize the actual model. The reason behind using only those degrees of freedom necessary to characterize a physical problem is that in using a personal computer, the user is limited to the maximum amount of storage available in the particular personal computer configuration to solve a problem. To allow a large problem to be run on a personal computer, certain algorithms are used to reduce the storage requirements.

The basic method of increasing the allowable problem size is a technique of reducing the storage requirements for the stiffness matrix. In general, the full stiffness matrix is not stored and is reduced during the solution process. In essence, the technique stores only the nonzero terms of the matrix. These nonzero terms are contained within a "band." The bandwidth of a structure is proportional to the maximum nodal difference among the connecting nodes for all of the elements in the model. In other words, the bandwidth can be calculated from the following equation:

$$BW = (\text{HIGH NODE NUMBER} - \text{LOW NODE NUMBER} + 1) \times \text{DOF} \tag{3.1}$$

The maximum value of this group of numbers defines the bandwidth of the structure.

All programs contain a node-renumbering algorithm that maps the user-defined node numbers to equivalent system numbers. The system node numbers allow the program to run more efficiently during the analysis. The user sees only the results at the nodes he or she defines.

By renumbering the nodes and by using the banded solution technique, the solution time is significantly reduced, as is the amount of disk storage required.

In the development of a finite element model, the most important item on which to focus is the bandwidth.

3.5.2 Mesh size

One of the more frequently asked questions concerns the generation and selection of the proper mesh size in order to analyze a problem. Each problem is different and there are no definite rules to develop the proper mesh size. Engineering judgment and intuition are called for to know where the regions of excessive stress or strain will be located. The problem geometry will dictate the areas where prominent changes in geometry occur, requiring a finer mesh in that particular area. Also, the personal computer configuration does influence the mesh size via the available storage capacity of the unit.

3.5.3 First runs with selected mesh

The user must review the output of his or her modeling efforts. This is especially true for the first solution obtained from a particular program. There are items to look for that will determine if the mesh size was too coarse or too fine for the particular application. Look for disproportionate stress level changes from node to node or plate to plate, and large adjacent node displacement differences. If a sign change occurs, this is an indication that the elements are too large to produce accurate results. If the mesh density can be doubled with a change of one percent in the results, either model is satisfactory. Although mesh density studies are time-consuming, this technique may prove worthy where many analyses are to be performed on similar components. The user must use caution and not trust the finite element program explicitly. Paying attention to the details will result in the finite element program answers being accurate.

3.5.4 Solution accuracy

A distinct concern of every engineer who uses the finite element method in solving problems is the accuracy of the solution. Again, the fundamental concept of finite element modeling is that any continuous quantity can be approximated by a discrete model composed of a set of piecewise continuous functions defined over a finite number of subdomains. The geometry of the problem under consideration is defined in the finite element approach by dividing the structure into subdivisions or elements. The elements are connected by nodes. As the number of nodes is increased (and, in turn, the number of elements is increased), a closer approximation to the real-world physical problem is achieved. This, however, leads to very lengthy calculation times on the personal computer. Defining the model accurately is the user's primary concern. Nodes should be defined at locations where changes of geometry or loading occur. Changes in geometry relate to thickness, material, and/or curvature. The accuracy in the locations of loads and restraints and the proper definition of the restraints are critical in obtaining a reasonable solution. Modeling of the geometry is perhaps the easiest area in which to improve the accuracy of the solution.

Every model is different. To verify the solution, a load case should be used for which an approximate answer can be obtained. One may wish to use, if a physical model is available, measured data. This will allow quick verification of the model's accuracy. In addition, the verified model can be used to examine "*what if*" options with confidence in the results.

3.5.5 Refining the mesh in critical areas

A straightforward check for accuracy of the model is to increase the number of elements by fifty percent and compare the results. If large changes in stresses or deflections occur, or if a sign change occurs, the original element size was too large to give accurate results. The following discussion focuses the selection of two commonly used plate elements and their influence on solution accuracy. A complete dissertation on all elements and their influence on the accuracy of the solution is beyond the scope of this text.

To study the plate element and the accuracy of a solution derived from the use of the plate element, two basic elements are selected: (1) the isoparametric quad element, and (2) the triangular element.

In any selection of the type of element to use, the engineer must carefully consider how the structure will react to the loads imposed on the structure.

There are two basic types of load-carrying mechanisms for plates. The first is membrane, and the second is bending. If the engineer feels that both types of forces are expected, the better element to use is the one that allows both load-carrying mechanisms. Experience has shown that better stress results are obtained with the quad element than with the triangular element. Triangular elements are best used where dictated by geometry. The triangular elements are often used for modeling transition regions between fine and coarse grids, for modeling irregular structures, for modeling warped surfaces, etc. When using triangular elements in a rectangular array of nodal points, the engineer can obtain the best results from an element pattern having alternating diagonal directions. Triangular two-dimensional solid elements are less accurate than equivalent-sized quadrilateral elements and should not be used in highly stressed areas. Quadrilateral plate elements should lie in an exact flat plane. Warped areas should be modeled with triangular elements. Warped quadrilateral elements will cause a loss of equilibrium since the element resisting stiffness is based on the element plane defined by the first three nodes. If the fourth node does not lie in the element plane, a moment imbalance results. If the model is one that has, for example, a thick shell, then one finds the thick plate variations of the elements the better ones to use for an analysis. Note that a plate is considered to be thick if the overall length-to-thickness ratio is less than four. For applications involving cylindrical coordinates, a thick plate should be used when the radius-to-thickness ratio is less than four.

3.5.6   Practical suggestions, limitations, and interfacing to mainframes

There are several additional items to consider in the modeling of a physical problem beyond the type of element to select. These are aspect ratio, flexibility of the physical structure, boundary conditions, thickness, and applied forces.

The aspect ratio of an element is the ratio of the length of a defined base to the height of the triangle. For the quad element, either regular or skewed, the ratio is the length of one side to an adjacent side. Triangular elements should have an aspect ratio between one and two. If the isoparametric quad element is used, the aspect ratio should be below two. Deflections vary greatly with theory if the aspect ratio is greater than two. Note that some special quad elements such as the Crisfield quad can have an aspect ratio up to ten without a loss of accuracy if membrane stresses are not significant.

The flexibility of a structure influences the number of elements to be used. For example, a very flexible structure to model, such as a flexible hose, would require many more elements than an equivalent (physically) metal hose or concrete duct. The engineer must pay more attention to the aspect ratio of the elements used. Experience can lead to the correct choice of appropriate constraints based on the model geometry and materials.

Boundary conditions become important in evaluating the results. Structures with boundaries having a degree of fixity between "simply supported" and "rigid" requires the use of experienced engineering judgment in the setting of the boundary conditions. Boundary conditions can be a source of singularities in a model. A singularity exists in a problem whenever an indeterminant or undefined solution is possible. Singularities can be caused by the following boundary related conditions.

1. Unconstrained structure. All structures should be constrained to prevent rigid body motions.

2. Unconstrained joints. Singularities can exist at a particular degree of freedom due to the element arrangements.

Thickness is of concern only for extremely thick shells. If the thick plate element is used, then the aspect ratio becomes an important parameter because of the shear deflection.

It is often difficult to define the types of forces, magnitude, location, and distribution. Besides properly defining the boundary conditions, selecting the forces to be applied and the eventual quantification of those forces may be one of the most difficult aspects of finite element modeling. Applied forces include, but are not limited to, static point, static distributed, transient, harmonic vibratory, and thermally induced loadings. That the forces are applied correctly from the standpoint of magnitude, distribution, direction, and frequency and phasing, if necessary, is imperative.

There are occasions where it is impossible to accurately model a given problem with the limitations of the personal computer. Many FEA programs allow interface to mainframe computers that have versions of codes such as ANSYS or NASTRAN.

The finite element program, ALGOR SuperSAP, used in this book allows the user to create files compatible with NASTRAN. Transfer of the node and element files occurs via communications software and a modem. Therefore, the PC-based finite element software is used as a preprocessor for the mainframe codes. This reduces the user's connect time to the mainframe during the model building process.

3.6 Preprocessing

Preprocessing is the initial part of any modeling process. This includes the areas discussed above for the selection of nodes, elements, loads, and constraints. Most FEA programs have a form of preprocessing, although that particular feature of the model may not be called out as such in a particular FEA program.

3.6.1   Assembling the model

At this point in the analysis of a particular problem, the user must generate the required input to model the physical problem. Nodes are selected to represent points on the boundaries, the surface(s), and any positions interior to the physical problem.

Other constants must be defined in order for the model to run. These include, but are not limited to, the material properties (modulus of elasticity, Poisson's ratio, material density, material conductivity, material coefficient of thermal expansion, etc.), the cross-sectional areas, the moment of inertia(s) (*Ixx, Iyy, Izz*), the plate thickness (if applicable), etc.

3.6.2   Defining constraints (restraints)

The definition of the proper restraints for a problem is, in most cases, not an easy or straightforward task. Often, this part of the analysis is not clearly defined and the user is required to rely on his or her best engineering judgment. If the user's assumptions are not correct, then the model may act as a rigid body. Normally, the FEA programs do not model the behavior of rigid bodies unless directed to consider this type of motion.

A restraint is applied to a node. This restraint restricts movement in a given direction. This will in turn require an output from the program of a reaction force at that node. Another term to become familiar with is *member end release*. A member end release is the removal of a force-carrying ability on the member, thus allowing movement.

Each unrestrained node has six degrees of freedom: *X, Y, Z* translation, and rotation about the *X, Y,* and *Z* axes. The entire structure also has six rigid body degrees of freedom. Since only a few structures have rigid body motion, boundary conditions must be applied in order to restrain rigid body motion.

Two-dimensional planar structures must have the out-of-plane degrees of freedom zeroed. A planar structure defined in the XY plane would have the X and Y rotations zeroed and the Z translation zeroed. In addition, at least three additional degrees of freedom in the XY plane must also be zeroed in order to constrain rigid body motion in the XY plane.

In a two-dimensional problem and considering the XY plane, a clamped boundary condition would have the X and Y translations equal to zero and the Z rotation set equal to zero. A knife edge support would have the X and Y translations set equal to zero and the Z rotation would not be set equal to zero. If the user were modeling, a "roller" type boundary condition would force the Y translation and Z rotation to be set equal to zero and allow the X translation to be free.

Three-dimensional structures have six rigid body degrees of freedom. At least six degrees of freedom need to be restrained to prevent rigid body motion. This can be accomplished by zeroing translations and rotations as dictated by the problem.

3.6.3 Defining loads

3.6.3.1 Nodal loading.   Each FEA program has the ability to apply forces at nodal locations in a model in order to simulate the actual structure loading. These loads (Fx, Fy, Fz, Mx, My, Mz, Temperature, Convective Boundaries, Heat Flows, and/or Heat Fluxes) are input at a given node in and about the respective global coordinate directions. The loads at these nodes can also be specified as applied nodal displacements.

3.6.3.2 Plate loading.   In some applications, the loading of a model may be determined by internal and/or external pressures. A typical example would be the modeling of a soft drink can shaken vigorously. A model of this system requires the input of the increased pressure internal to the soft drink can and the outside ambient pressure. Some FEA models allow the input of plate pressures. The plate pressure input is converted to an equivalent nodal loading by taking into account the shape functions and area of the plate associated with a particular node. If the user has a program that does not contain the option of specifying this particular type of loading, a first cut at this type of analysis would be to assume that all nodes are loaded equally.

3.6.4   Defining nodal weights

3.6.4.1   FEA program-generated weights.   In any structure, weight is involved. FEA programs will automatically assign a weight to each node. In general this weight is the total structural weight divided by the number of nodes if the model of the structure has uniform material properties. For those models constructed with elements of different materials, the appropriate nodal weight will be assigned by the FEA program. The user should be aware that some programs do not consider the weight of the structure in the static stress analysis portions of the program.

3.6.4.2   User-generated weights.   There may be occasions when the user (engineer, programmer) must apply additional nodal weights to compensate for real-world situations that will have a noticeable effect on the solution. This option allows a closer approximation to proper simulation without the need for additional modeling such as the use of more nodes and elements. Most FEA programs allow the user to specify these additional weights either instead of or in addition to the FEA program-generated weights.

3.7   Executing the Model

With all parts of the model defined (nodes, elements, restraints, and loadings), the analysis phase of the model is ready to begin. Stresses, deflections, temperatures, pressures, velocities, and vibration modes can be determined. The solution of the equations is accomplished in one of two ways. The first is to use a *wavefront* solution. In general, node numbers are arbitrary and certain nodes are ignored in the analysis. In bandwith-based programs, nodes must be numbered and sequenced. This operation is usually done with certain mathematical algorithms. In the analysis phase, there are certain occurrences basic to any FEA model that must happen to effect the solution to any particular problem. The three following sections address the sequence of events necessary to execute a static stress model, a vibration model, and a thermal model.

The first step for any of the three types of analyses is to perform a check of the preprocessing activities. All FEA programs include routines to check the geometry of the model. The areas checked include nodes (check for duplication), elements (check for type, proper connectivity, all specified constants), and material properties (check for modulus of elasticity, density, Poisson's ratio, conductivity, thermal expansion coefficient, etc.).

The next step is to renumber the nodes for the internal system of equations. Renumbering is beneficial from the standpoint of reducing used disk space and run time for the type of analysis to be performed. As discussed earlier, reducing the bandwidth of the stiffness matrix is imperative. This is an important point that cannot be overstated. As the user becomes familiar with the FEA program being used, he or she can construct better models that reflect user-inspired small bandwidths.

After renumbering the nodes, the next step is to calculate the element stiffness matrices (or element conductivities, if performing a heat transfer analysis) and assemble these into the global stiffness file. All FEA programs generally use the same technique to accomplish this task. The difference lies in the algorithm used to perform the stiffness assembly task.

3.7.1   Static stress analysis

After the model check, node renumbering, and element
stiffness assembly, the next step is to solve the matrix equations for the nodal displacements. For the static stress analysis, the form of the equation to solve is

$$[K] [U] = [F] \qquad (3.2)$$

where: $[K]$ = Stiffness Matrix
$\quad\quad$ $[U]$ = Nodal Displacements
and
$\quad\quad$ $[F]$ = Nodal Forces.

The solution to this set of equations is accomplished by
the Gauss elimination technique or some derivative of the
technique.

The final step in this process is to calculate the stress
associated with the deflected shape. The details of this calculation are beyond the scope of this book. It is sufficient to
say that the user should refer to most any mechanics of
materials text to review or develop an understanding of the
stress displacement relationship. Computer implementation from that point is straightforward.

Stresses are calculated at the element centroids and are
averaged for postprocessing at the nodes.

In a static analysis, there are certain points to remember
in the development of a consistent and error-free model.

3.7.1.1   Nodes.   Increasing the number of nodes used for
the model increases the accuracy of the model. However,
solution and debug time increase. Node point spacing is not
required to be uniform. It is recommended that a fine mesh
be used in regions of high stress or where the stresses are
changing rapidly. A coarse mesh should be used in areas of
low stress or where the stresses are nearly constant.
Transition in mesh density should be smooth.

3.7.1.2   Loads.   Loads can be applied to each node point in
a model if desired. At times, concentrated loads should be

applied across a minimum of two nodes for continuous structures. This allows the entire element to be loaded, rather than just one point of the element.

3.7.2 Thermal analysis

A thermal analysis is quite similar to the static stress analysis discussed in Sec. 3.7.1. The stiffness matrix is replaced by a conductivity matrix with the solution technique similar to that for the static case. The theory behind the conductivity matrix is that the conductivity of the item must be discretized as the stiffness in the static stress analysis. The output from this analysis is all nodal temperatures.

The FEA heat transfer calculations that output the nodal temperatures are very useful by themselves. However, most users are interested in the ability of an FEA program to calculate thermal stresses. Thermally induced stresses and displacements are caused by the following conditions:

1. Differential expansion of dissimilar materials in the same structure or system

2. Uneven or transient heat flux applied to the system

3. A rigid constraint

The use of FEA for thermal analysis will generally fall into one of three categories. These are *Steady State Heat Transfer, Transient Heat Transfer,* and *Thermally Induced Displacements and Stresses.*

Steady state heat transfer equations for conduction are basically the Laplace equation for each node point. The basic equation, in matrix form, is given by

$$[C]\,[T] = [Q] \tag{3.3}$$

where $[C]$ is the global conductivity matrix, $[T]$ is the nodal temperatures, and $[Q]$ is the nodal heat flux vector. The heat flux vector is resolved into nodal loadings. If convection boundary conditions are specified, the set of matrix equations is given by

$$[C][T] = [Q] + [h(T)][T] \tag{3.4}$$

where $[h(T)]$ is the set of convection coefficients.

Transient heat conduction is expressed by the following equation:

$$\frac{\partial}{\partial x_i}\left(k\frac{\partial T}{\partial x_i}\right) = \rho c_p \frac{\partial T}{\partial t} \tag{3.5}$$

In the above equation, ρ is the mass density, c_p is the material specific heat, and k is the material conductivity. Programs that provide transient thermal capabilities contain specific time-marching algorithms in order to implement a time-dependent FEA solution formulation.

For the thermally induced displacements and stresses, nodal temperatures are included in the static displacement and stress solution as a set of nodal loadings as follows:

$$[K][D] = [F] + [F_T(T)] \tag{3.6}$$

where $[K]$ is the stiffness matrix, $[D]$ is the displacement vector, $[F]$ are the nodal loadings other than thermally induced, and $[F_T(T)]$ is the nodal loading vector due to temperature considerations.

3.7.3   Vibration (dynamic) analysis

A vibration analysis is more complicated than the static stress case. Element mass matrices must be calculated. For natural frequency analysis, the mass of the object must be discretized, as must the stiffness of the element. The elements use a lumped mass technique such that the total mass of each element is distributed among its nodes with a method similar to the resolution of the system stiffness to the stiffness between the nodal points.

The dynamic response of a system is a function of three parameters:

1. System stiffness and mass properties

2. Certain dynamic forces driven by amplitude, frequency, and relative phase angle

3. Damping due to material damping, viscous damping, and Coulomb damping

Some programs do allow the user to specify an overall damping number or coefficient.

After assembling the stiffness and mass matrices, the set of equations is solved for the eigenvalues and eigenvectors. The natural frequencies come from the eigenvalues, and the deflected mode shapes relate to the eigenvectors. The solution of the set of equations requires an iterative solution. Each FEA program effects a solution in a different manner. Due to the number of calculations and storage requirements for a dynamic analysis, some programs rely on condensation techniques to decrease the size of the global problem and reduce the dynamic degrees of freedom sufficient to characterize the dynamic response of the structure. A popular form of dynamic condensation is Guyan Reduction. Using this procedure requires the reduction of the original stiffness and mass matrices to a specified number of dynamic degrees of freedom. Stiffness terms are reduced independently of the mass terms. The mass terms are redistributed according to the reduced stiffness matrix.

Another method used to determine the natural frequencies is called subspace iteration. This method does not require the problem size to be reduced. The procedure begins by solving for the lowest modes first. Upon completion of this step, the mode shapes as well as the system stresses are calculated.

In equation form,

$$[K]\,[D] = [M]\,[D]\,[W]^2 \tag{3.7}$$

where $[D]$ = displacement matrix and $[W]$ = diagonal matrix containing the eigenvalues.

3.8   Postprocessing

The postprocessing of the data generated from an analysis is an important phase of any problem. The postprocessors associated with the FEA programs reviewed in this text are quite good and offer much insight into the program-generated results.

In addition to organization of the results in an orderly fashion, the FEA postprocessors offer graphical output to the monitor, printer, and/or plotter.

In the case of the static stress and thermal calculations, most programs offer the capability of plotting the stress/temperature contours on the undeflected as well as the deflected model.

Output from a vibration analysis is usually given as the deflected shape with the accelerations and stresses associated with the frequency analysis. A particularly useful form of the output is the animation of the calculated mode shapes. Animation helps the user visualize exactly what is happening to the item modeled, and allows the user to assess the reality of the modeling process.

3.9   Design Optimization

After the completion of a structural analysis, a question often asked is whether or not a part is overdesigned from either a weight or cost standpoint. Optimized designs are those designs which meet all of the strength requirements while minimizing such factors as cost and weight.

In order to achieve the optimum structural design, the following four criteria must be satisfied:

1. The design satisfies all necessary standard engineering practices.

2. All stresses are below the allowable stresses.

3. All vibration, natural frequency, and deflections (both static and dynamic) are within specifications.

4. The structural weight and/or cost are minimized.

Structural optimization is an iterative process. This process involves the repeated analysis of a structural system while changing the design variable values based on the previous analysis result. The design variables may be any selected by the user. Both upper and lower bounds must be selected by the user. The bounds may be physical constraints or behavioral constraints. Physical constraints could include minimum or maximum allowable cross-sectional areas. Behavioral constraints place a limit on stress or perhaps one of the fundamental structural resonant points.

The goal of the optimization process is to produce a design where each element is at or near a fully stressed state for at least one of the specified load cases. However, the fully stressed state does not necessarily represent the minimum weight goal. In turn, the minimum weight goal could lead to a structure that is unstable from a buckling or resonant frequency standpoint.

4

Introduction to Using the ALGOR Interactive Systems, Inc. FEA Package and Verification Problems

4.1 Introduction

The purpose of this chapter is to examine the application of FEA to five problems with verifiable results. A simple problem introduces the basics using a three-dimensional truss. The details of imputing data and using the program efficiently are best left to the ALGOR manuals listed in the reference section. These manuals detail how to use the program to build and prepare the models for analysis.

4.2 Three-dimensional Truss

The problem illustrated in this section is shown in Fig. 4.1 for the determination of stresses in each member. Therefore it is necessary to determine the tension in each of the truss members.

Point 4 is chosen as a free body and is subjected to four forces. The unknowns are the tensions in the members defined by endpoints 4 to 1, 4 to 2, and 4 to 3.

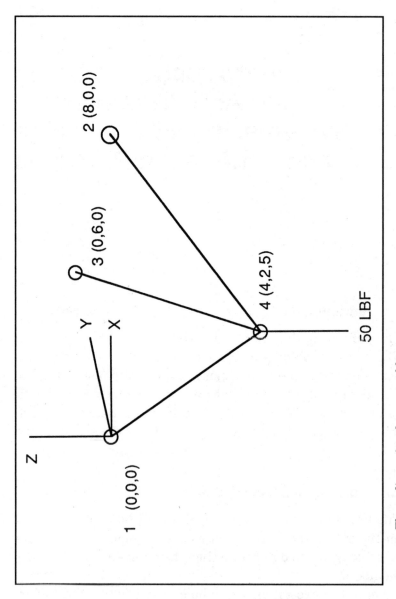

Figure 4.1 Three-dimensional truss problem.

It is necessary to determine the components and magnitudes of the vectors α_{41}, α_{42}, and α_{43}.

$$\alpha_{41} = 4i + 6j - 2k$$
$$\alpha_{42} = -4i + 6j - 2k \qquad (4.1)$$
$$\alpha_{43} = -4i + 6j + 4k$$

The unit vectors corresponding to α_{nm} are

$$\lambda_{41} = +.5345i + .8018j - .2673k$$
$$\lambda_{42} = -.5345i + .8018j - .2673k \qquad (4.2)$$
$$\lambda_{43} = -.4851i + .7276j + .4851k$$

The tension in each cable is written as follows:

$$T_{41} = +.5345T_{41_i} + .8081T_{41_j} - .2673T_{41_k}$$
$$T_{42} = -.5345T_{42_i} + .8018T_{42_j} - .2673T_{42_k} \qquad (4.3)$$
$$T_{43} = -.4851T_{43_i} + .7276T_{43_j} + .4851T_{43_k}$$

Since point 4 is in equilibrium, the sum of the forces is zero, or

$$T_{41} + T_{42} + T_{43} = 0. \qquad (4.4)$$

Substituting the above expressions for T_{ij} and setting the coefficients of i, j, and k equal to zero results in three equations for the three unknowns. The solution to the set of equations can easily be obtained by Gaussian elimination resulting in the following solution:

$$T_{41} = 10.391 \; lb_f$$
$$T_{42} = 31.180 \; lb_f \qquad (4.5)$$
$$T_{43} = 22.906 \; lb_f$$

The stress in each truss member is determined by dividing the tension in each member by the cross-sectional area. For this example problem, the cross-sectional area is 1 in^2.

4.3   Finite Element Solution

The FEA solution to the above problem is quite straightforward. All steps used to arrive at a solution are given and can be run on any of the FEA programs the user may have.

4.3.1   Nodes

After entering the program and getting into the preprocessing portion of the program, the first step is to input the nodes and the node coordinates. For this particular problem, the nodes and node coordinates are given below.

Node	x (in)	y (in)	z (in)
1	0	0	0
2	96	0	0
3	0	72	0
4	48	24	60

4.3.2   Material properties

The only material properties needed for a three-dimensional truss element static analysis is the modulus of elasticity, density, and Poisson's ratio. Assume that the truss element is steel. Therefore, E is equal to 30.0E06 lb/in^2, the density is equal to 0.283 lb/in^3, and Poisson's ratio is 0.3.

4.3.3   Real constant(s)

Beam and Truss elements require the input of various physical data such as cross-sectional area and various moments of inertia. The three-dimensional truss requires only the cross-sectional area. The value is 1.0 in^2.

4.3.4   Truss connectivity

The following list is the element connectivity. Included are truss length from node to node, and the direction cosines (as determined from the problem definition) for each beam.

| Beam | Nodes | | | Direction Cosines | | |
No.	From	To	Length	x	y	z
1	1	4	7.483	−0.15	0.96	−0.22
2	2	4	7.483	0.15	0.96	−0.22
3	3	4	8.246	0.27	0.87	0.40

4.3.5 Restraints

In any problem the determination of the proper restraint to apply at each node is very important. In this problem the use of a three-dimensional truss element dictates that any rotations about the x, y, or z axes is prohibited. Only translations in each coordinate direction are allowed. The movements in each coordinate direction for nodes 1, 2, and 3 will be restrained; that is, displacements in the x, y, and z directions for nodes 1, 2, and 3 will be equal to zero.

4.3.6 Loads

The only load for this problem is one of 50 pounds applied at node 4 in the $-z$ direction.

4.3.7 Solution

After imputing all of the above information, the next step is to check the input for consistency. Most programs have subroutines to check if all information required to perform an analysis has been imputed to the program. Following this step is the requirement to renumber the nodes. This is internal to the program and allows an improvement in the solution efficiency. The following printout illustrates the screens and solution steps for the truss problem given above.

1**** Algor (c) Linear Stress Analysis - SSAP0H 9/07/90, Ver. 9.03/387

```
    DATE: FEBRUARY 2,1991
    TIME: 01:56 PM
    INPUT FILE.............FOUR1
----------------------------------------------

    FOUR1.
```

1**** CONTROL INFORMATION

number of node points	(NUMNP)	=	10
number of element types	(NELTYP)	=	1
number of load cases	(LL)	=	1
number of frequencies	(NF)	=	0
geometric stiffness flag	(GEOSTF)	=	0
analysis type code	(NDYN)	=	0
solution mode	(MODEX)	=	0
equations per block	(KEQB)	=	0
weight and c.g. flag	(IWTCG)	=	0
bandwidth minimization flag	(MINBND)	=	0
gravitational constant	(GRAV)	=	3.8640E+02

bandwidth minimization specified

1**** NODAL DATA

NODE	BOUNDARY CONDITION CODES						NODAL POINT COORDINATES			
NO.	DX	DY	DZ	RX	RY	RZ	X	Y	Z	T
1	1	1	1	1	1	1	0.000E+00	0.000E+00	-1.000E+14	0.000E+00
2	1	1	1	1	1	1	0.000E+00	-1.000E+14	0.000E+00	0.000E+00
3	1	1	1	1	1	1	-1.000E+14	0.000E+00	0.000E+00	0.000E+00
4	1	1	1	1	1	1	0.000E+00	0.000E+00	0.000E+00	0.000E+00
5	1	1	1	1	1	1	9.600E+01	0.000E+00	0.000E+00	0.000E+00
6	1	1	1	1	1	1	1.000E+14	0.000E+00	0.000E+00	0.000E+00
7	1	1	1	1	1	1	0.000E+00	7.200E+01	0.000E+00	0.000E+00
8	1	1	1	1	1	1	0.000E+00	1.000E+14	0.000E+00	0.000E+00
9	0	0	0	1	1	1	4.800E+01	2.400E+01	6.000E+01	0.000E+00
10	1	1	1	1	1	1	0.000E+00	0.000E+00	1.000E+14	0.000E+00

**** PRINT OF EQUATION NUMBERS SUPPRESSED

1**** TRUSS ELEMENTS

number of truss members =   3
number of diff members =   1

1**** MATERIAL/AREA DATA

INDEX	E	ALPHA	MASS DENSITY	AREA	WEIGHT DENSITY
1	3.0000E+07	0.0000E+00	7.3240E-04	1.0000E+00	2.8300E-01

1**** ELEMENT LOAD MULTIPLIERS

CASE A   CASE B   CASE C   CASE D
---------- ---------- ---------- ----------
X-DIR 0.000E+00 0.000E+00 0.000E+00 0.000E+00
Y-DIR 0.000E+00 0.000E+00 0.000E+00 0.000E+00
Z-DIR 0.000E+00 0.000E+00 0.000E+00 0.000E+00
TEMP 0.000E+00 0.000E+00 0.000E+00 0.000E+00

1**** ELEMENT CONNECTIVITY DATA

ELEMENT NUMBER	NODE I	NODE J	MAT'L INDEX	TEMP
1	4	9	1	.00
2	7	9	1	.00
3	5	9	1	.00

1**** BANDWIDTH MINIMIZATION

minbnd (bandwidth control parameter) =   1
**** MINIMIZER DID NOT NEED TO REDUCE BANDWIDTH

**** EQUATION PARAMETERS
total number of equations   =   3
bandwidth   =   3
number of equations in a block =   3
number of blocks   =   1
blocking memory (kilobytes) =   4762
available memory (kilobytes) =   4762

**** Hard disk file size information for processor:

Available hard disk space on drive =   26.485 megabytes
Estimated required hard disk space =   .400 megabytes

1**** NODAL LOADS (STATIC) OR MASSES (DYNAMIC)

NODE NUMBER	LOAD CASE	X-AXIS FORCE	Y-AXIS FORCE	Z-AXIS FORCE	X-AXIS MOMENT	Y-AXIS MOMENT	Z-AXIS MOMENT
9	1	0.000E+00	0.000E+00	-5.000E+01	0.000E+00	0.000E+00	0.000E+00

1**** ELEMENT LOAD MULTIPLIERS

load case   case A   case B   case C   case D
---------- ---------- ---------- ---------- ----------

1   0.000E+00 0.000E+00 1.000E+00 1.000E+00

1**** STIFFNESS MATRIX PARAMETERS

minimum non-zero diagonal element =   1.7512E+05
maximum diagonal element   =   6.1525E+05
maximum/minimum   =   3.5132E+00
average diagonal element   =   3.9471E+05
density of the matrix   =   6.6667E+01

1**** TEMPORARY FILE STORAGE (MEGABYTES)

UNIT NO. 7 :   .042
UNIT NO. 8 :   .048
UNIT NO. 9 :   .042
UNIT NO. 10 :   .000
UNIT NO. 11 :   .000
UNIT NO. 12 :   .000
UNIT NO. 13 :   .000
UNIT NO. 14 :   .000
UNIT NO. 15 :   .042
UNIT NO. 17 :   .000

TOTAL   :   .175

1**** End of file

****
Information for processor
****

1**** Algor (c) Linear Stress Analysis - SSAP0H 9/07/90, Ver. 9.03/387

DATE: FEBRUARY 2,1991
TIME: 01:56 PM
INPUT FILE.............FOUR1

**** Hard disk file size information for processor:
No. of equations   (NEQ) =   3
Bandwidth   (MBAND) =   3
No. of eqs./block (NEQB) =   3
No. of blocks   (NBLOCK) =   1

Available hard disk space on   drive =   26.485 megabytes
Estimated required hard disk space =   .400 megabytes

**** Table for actual hard disk space used:
FOUR1.T12   =   42.240 kilobytes
FOUR1.T8   =   48.000 kilobytes
FOUR1.T9   =   42.240 kilobytes

```
FOUR1.T15    =        .000 kilobytes
FOUR1.T14    =        .184 kilobytes
FOUR1.T7     =        .032 kilobytes
FOUR1.T13    =        .000 kilobytes
FOUR1.T11    =        .000 kilobytes
FOUR1.T10    =      42.240 kilobytes
FOUR1.T17    =        .000 kilobytes
FOUR1.L      =       7.301 kilobytes
FOUR1.DO     =        .528 kilobytes
FOUR1.SO     =        .000 kilobytes
```

Total actual hard disk space used   =   .183 megabytes

Sub-total elapsed time   =   .501 minutes

****
Information for postprocessor
****

1**** Algor (c) FEA Stress Processor MKNSOH Rel 07/06/90 Ver 1.02/387

DATE: FEBRUARY 2,1991
TIME: 01:56 PM
INPUT FILE.............FOUR1

**** Hard disk file size information for postprocessor:
```
FOUR1.SON    =        .624 kilobytes
FOUR1.S      =        .770 kilobytes
FOUR1.NSO    =        .054 kilobytes
```

Total postprocessing disk space used =   .001 megabytes

MKNSO   elapsed time   =   .095 minutes

The TOTAL elapsed time   =   .596 minutes

1**** Algor (c) FEA Stress Processor MKNSOH Rel 07/06/90 Ver 1.02/387

DATE: FEBRUARY 2,1991
TIME: 01:56 PM
INPUT FILE.............FOUR1

**** 3-D truss elements:

Number of elements   =   3
Number of materials   =   1

**** Nodal stresses for 3-D truss elements:

```
El. # LC ND   Stress      Force
----- --- -- ---------- ----------
   1  1 I  1.016E+01  1.118E+01
   1  1 J -1.016E+01 -1.118E+01

   2  1 I  2.288E+01  2.517E+01
   2  1 J -2.288E+01 -2.517E+01

   3  1 I  3.049E+01  3.354E+01
   3  1 J -3.049E+01 -3.354E+01
```

1**** End of file

Figure 4.2 illustrates the results obtained from the finite element analysis outlined above.

This technique of using a simple example is often quite useful for checking the accuracy of the model proposed, and is useful for understanding the basics and principles of the application of Finite Element Analysis to common textbook problems. This in no way suggests not studying or understanding the principles involved. FEA is simply an analytical tool that allows solutions to complex problems.

4.4   Additional Verification Problems

Many typical engineering problems can be found in any number of texts from the various engineering fields. The stress problems presented below are from the classic text of Roark and Young's *Formulas for Stress and Strain, 5th Ed.*[23] These problems are easily modeled in the ALGOR program. The formulas from the text are given, and the ALGOR solution is given for comparison.

4.4.1   Thick-walled cylinder with uniform internal pressure

This problem is a thick-walled cylindrical pressure vessel with a uniform internal radial pressure. All longitudinal pressure is zero or externally balanced. Assume that the material properties are that of steel and that the cylinder is

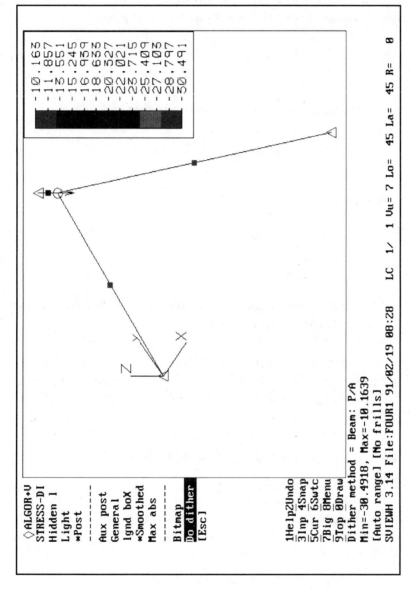

Figure 4.2 Stress analysis results—three-dimensional truss problem.

10 inches in outside diameter and 7 inches internal diameter. The internal pressure is 10,000 psi.

Roark and Young give the following equations for the normal hoop or tangential stress at a radius r as

$$\sigma_2 = \left[\frac{qb^2}{r^2}\right]\left[\frac{a^2+b^2}{a^2-b^2}\right] \qquad (4.6)$$

and the radial displacement at the inside radius as

$$\delta r = \left[\frac{qb}{E}\right]\left[\frac{a^2+b^2}{a^2-b^2}+n\right] \qquad (4.7)$$

where $a = 10''$
$\qquad b = 7''$
$\qquad q = 10{,}000$ psi
$\qquad n = .3$
$\qquad E = 3\text{E} + 7$.

The calculated values for σ_2 and δr are 29,216 psi and .007517 in, respectively.

Figures 4.3 and 4.4 show the calculated results from the stress analysis. The percent differences in the ALGOR calculated results and the Roark and Young calculated results for the hoop stress and deflection are 0.22 percent and 0.123 percent, respectively.

4.4.2   Temperature induced stress in thick-walled cylinder

The problem discussed in Sec. 4.4.1 is again used to illustrate the program capability to accurately predict temperature-induced stresses. From Roark and Young, the outer surface stress is given by

$$\sigma_{t_{outer}} = \frac{\Delta T \alpha E}{2(1-\upsilon)\log_e \dfrac{c}{b}}\left(1 - \frac{2b^2}{c^2-b^2}\log_e\frac{c}{b}\right). \qquad (4.8)$$

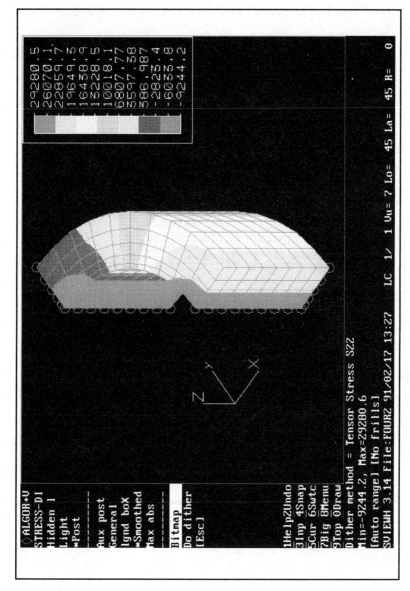

Figure 4.3 Stress distribution for 10,000 psi loading.

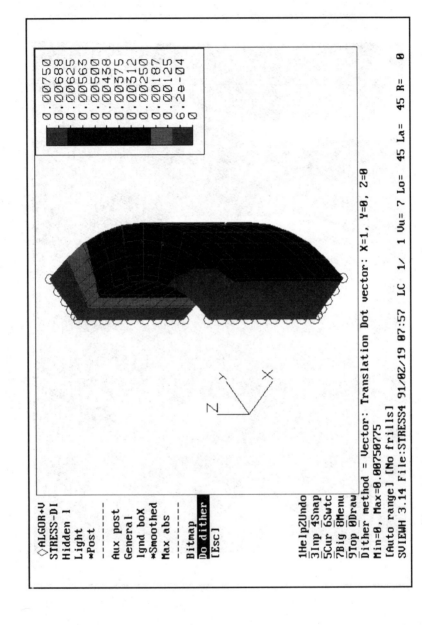

Figure 4.4 Deflection plot.

and the inner surface stress is given by

$$\sigma_{t_{inner}} = \frac{\Delta T \alpha E}{2(1-\upsilon)\log_e \dfrac{c}{b}}\left(1 - \frac{2c^2}{c^2-b^2}\log_e \frac{c}{b}\right) \tag{4.9}$$

Using the same physical model, temperatures of 0°F on the outer surface and 10°F on the inner surface are placed on the model. Using the steady-state thermal module of the ALGOR package, the temperature distribution for the fixed gradient across the cylinder wall is calculated. Note that no surface convection or radiation is involved. The results of this analysis are presented in Fig. 4.5. Using a module called ADVANCE in the ALGOR package transfers the calculated nodal temperatures to the physically equivalent stress model. The linear stress module is used to determine the stress levels imposed by the temperature difference across the wall. Results of the stress analysis are shown in Fig. 4.6.

The inner and outer FEA stress predictions vary from the Roark and Young calculations by a very few percent.

4.4.3 Lid-driven cavity flow

Even in manufacturing engineering there are problems involving the prediction of flow patterns in and around objects, as will be shown later in the book. The example problem that follows is one in which almost all incompressible flow codes are validated. This problem is that of lid-driven cavity flow for a Reynolds number of 400.[19]

An incompressible viscous fluid is trapped in a square two-dimensional 1×1 box. The top wall moves at a constant velocity of 1, thereby setting the fluid in the box in motion.

The following variables are defined:

Reynolds number Re = 1/v

Kinematic viscosity v = μ/ρ

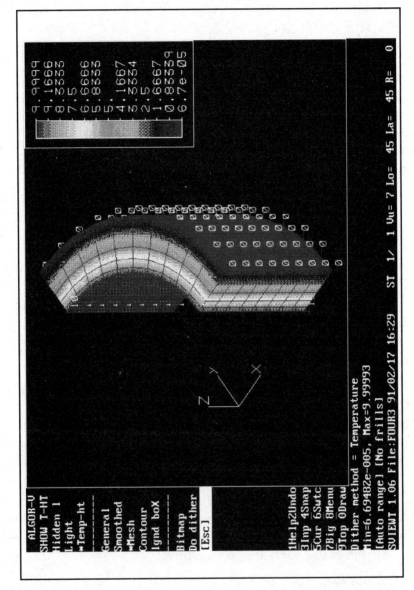

Figure 4.5 Temperature distribution.

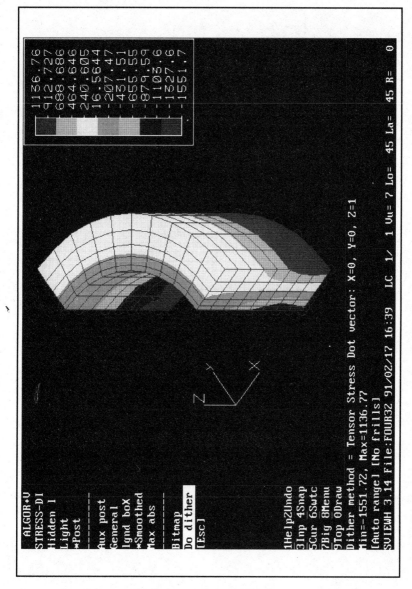

Figure 4.6 Temperature-induced stress distribution.

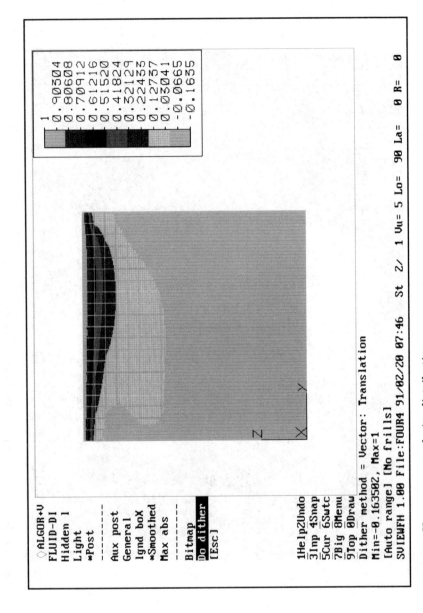

Figure 4.7 Y-component velocity distribution.

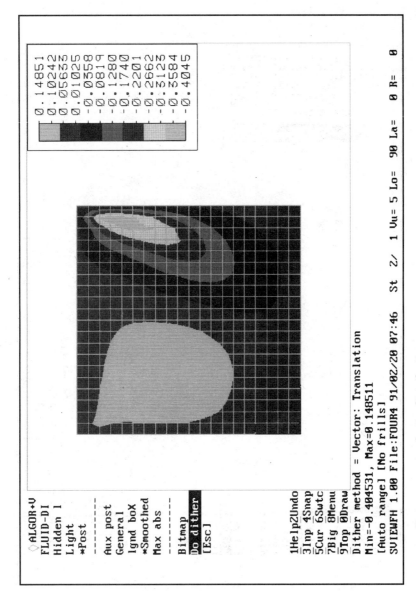

Figure 4.8 Z-component velocity distribution.

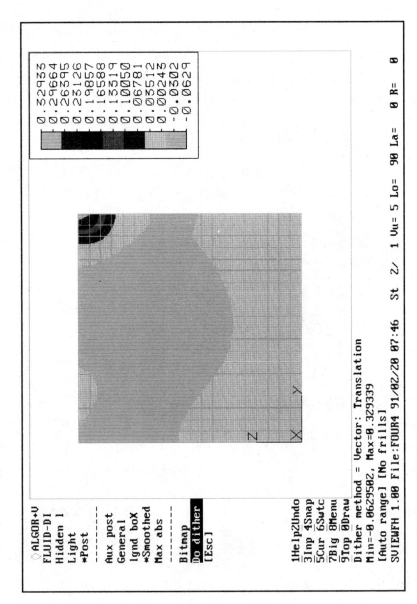

Figure 4.9 Pressure distribution.

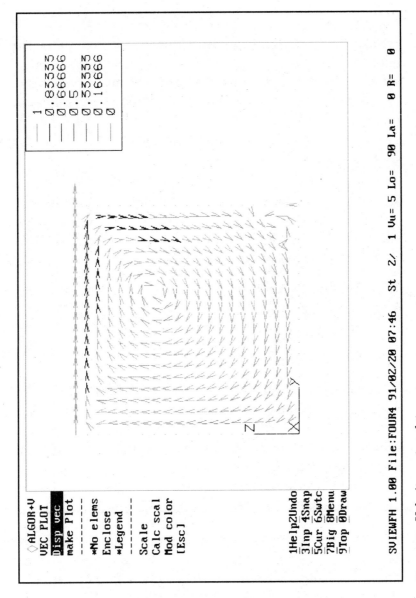

Figure 4.10 Velocity vector plot.

Dynamic viscosity μ

Mass density $\rho = 1$

A 20×20 mesh is constructed with $\mu = 0.0025$.

Figures 4.7, 4.8, 4.9, and 4.10 show the results of the analysis for the y-component of velocity, z-component of velocity, pressure distribution, and velocity vector plot, respectively. Results from this analysis match the reference material almost exactly.

4.4.4   A nonlinear analysis example

The ALGOR software contains modules for nonlinear analysis. This type of problem could include material nonlinearities or large deformation problems. Current modules include static analysis, dynamic analysis, and time history analysis.

Four analysis formulations are possible:

1. Linear Elastic Analysis

2. Materially Nonlinear Only Analysis

3. Total Lagrangian Formulation

4. Updated Lagrangian Formulation

Linear elastic analysis precludes any nonlinearities. The materially nonlinear only analysis assumes small strains and deformations while using a material with nonlinear properties. The total Lagrangian and the updated Lagrangian formulations can include both large deformations and material nonlinearities.

The problem presented herein is taken from the Verification Section of the AccuPak release notes. A bell-shaped rubber sheet is under a loading of 111.493 psi at the end surface. Figure 4.11 shows the finite element model. The rubber material is of the Mooney-Rivilin type with constants of $C1 = 21.605$ psi and $C2 = 15.7465$ psi. Sheet thickness is 0.125 in. Only half of the sheet was modeled (due to symmetry) and the two-dimensional plane stress elements

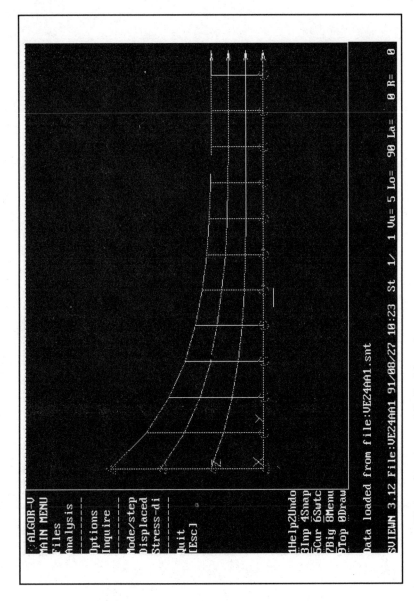

Figure 4.11 Finite element model of rubber sheet.

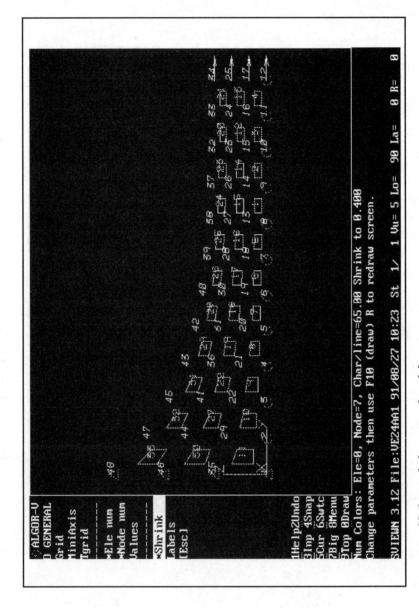

Figure 4.12 Nodes and elements of model.

92

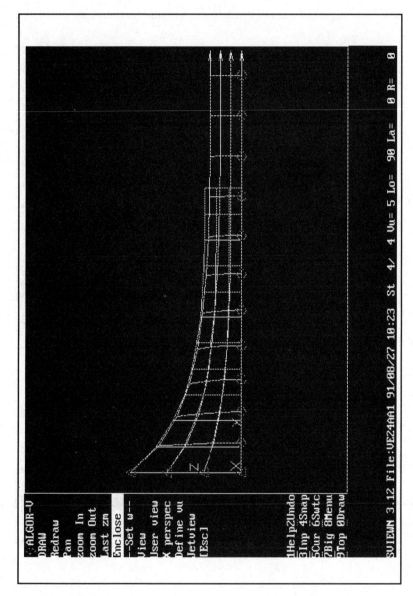

Figure 4.13 Deformed shape at load step four.

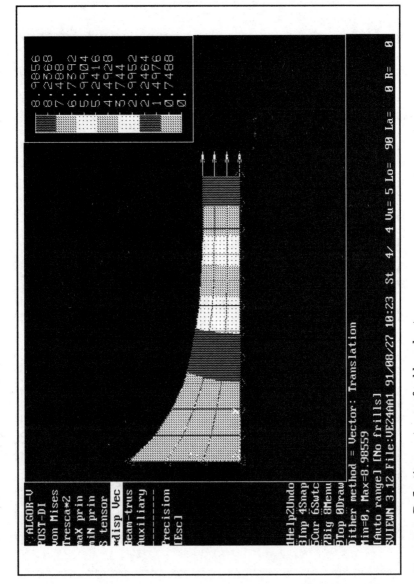

Figure 4.14 Deflection contours of rubber sheet.

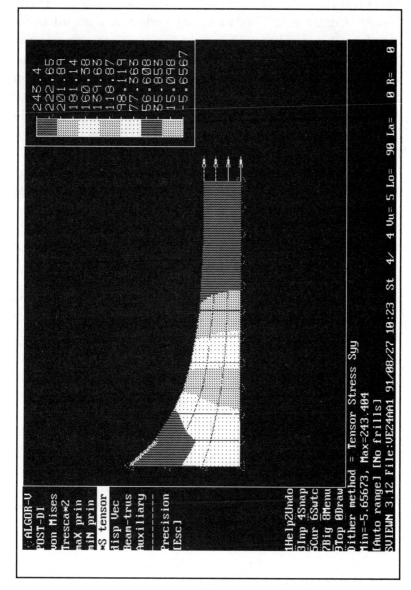

Figure 4.15 Stress values in rubber sheet.

were used. The uniformly distributed load of 111.493 psi
was converted to an equivalent concentrated load pattern.
The Total Lagrangian formulation was used to arrive at the
results. Figure 4.12 illustrates the node and element numbers for reference later.

Figure 4.13 shows an example of the deformed shape at
step four in the analysis. Figure 4.14 shows the deflection
plot. From the original reference, the results agree quite
well. The maximum predicted elongation is 8.973 in while
the published results indicate approximately 8.31 in.
Results are similar for each step in the analysis.
Improvements in mesh density could result in closer
approximations. Figure 4.15 illustrates the stress values at
the highest loading of 111.493 psi.

5

Applications of Finite Element Methods

5.1 Introduction

In this text I have tried to present examples that are useable. In addition, examples will be presented in the following chapters that illustrate how the FEA programs are used in practice. As the chapters have progressed, the material has gotten progressively more difficult but has been building on previous material. This chapter shows what is involved in a finite element program and provides a break between the overview given in Chap. 3 and the practical applications given in the remaining chapters.

This chapter attempts, in a very cursory manner, to give some background and insight into the finite element method through use of computer programs. It will be seen that finite element computations consist mainly of handling matrices and vectors. These operations include adding, multiplication, transposing, etc., much of which is covered in the typical undergraduate and graduate math courses. The programs given are written in FORTRAN and can be translated to the other languages as necessary.

The programs or subroutines given are not the most elegant or efficient, but they do illustrate how the method works.

5.2   What Is the Finite Element Method?

The finite element method is a technique for solving partial differential equations by first discretizing the equations describing the problem in their space dimensions. The discretization is carried out over small regions of arbitrary shapes. This results in matrices that relate the input at specified points to the output at all points in the domain. To solve the equations over large regions, the matrix equations for the smaller regions are usually summed node by node. This will result in the development of the global matrix equations. Many references are available that describe this technique. Perhaps the most notable is the text by Zienkiewicz and Taylor.[22]

We will set up a simple application for the example in this chapter: the plate element. The techniques are applicable to all of the more complex elements.

The last two decades have seen an explosive growth in computer technology. This growth does not show signs of abating. It appears that in the next few years this growth will prove to be even greater, particularly with the introduction of novel forms of architecture (e.g., vector, array, and multiprocessor systems). The computer hardware developments are most noticeable at the two extremes of the spectrum. At one end the massive computer systems, usually referred to as supersystems, such as CRAY-2 and CDC CYBER 205, have radically different and novel architectures that result in high performance [computational speed on the order of 100 million floating point operations per second (100 MFLOPS) or more]. At the other end of the spectrum are various types of minicomputers, engineering workstations (microcomputers), microprocessors, and handheld computers.

The introduction of these new computing systems has made a strong impact on finite element technology. The supersystems have made possible new levels of sophistication in finite element modeling as well as in problem depth and scope which were not possible a short while ago. These platforms have allowed development of finite element analysis software that allows the user to realize the full potential of the supersystems in finite element computation using special parallel numerical algorithms and efficient programming strategies.

The small, low-cost computer systems (minis, micros, and engineering workstations) provide a high level of interaction and free the finite element analysts from the constraints often imposed by large central computation centers.

Computer-aided design and computer-aided manufacturing (CAD/CAM) are terms that have gained prominence in the late 1970s and early 1980s to describe the modern engineering process in which virtually all stages of the process are unified through a computerized information repository and data flow. Among the stages of engineering usually included in the CAD/CAM rubric are conceptual design of overall systems, technical analysis and design of various subsystems, resolution of subsystem-integration difficulties, production of manufacturing or construction drawings and documents, programming of production, and generation of manufacturing control information. The unified computer-assisted process encompasses the development and evaluation of alternative designs and the capability for redesigning the subsystems or the overall system. The intent is to maximize both engineering creativity and productivity by providing the engineer with all the necessary information so that decisions and judgments can be made efficiently.

CAD/CAM systems are widely utilized in both research and industrial settings, although not as extensively as might be expected within some segments of engineering and industry. It is clear that further advances will occur as several developments in hardware and software are increasingly combined to create an appropriate environment. Three chief developments are networking, which enables the sharing and rapid transfer of large volumes of data; engineering workstations, which provide individual designers access to data and computing tools; and integral interactive-graphics capabilities, which present the opportunity for efficient control and manipulation of the complex engineering process. The obvious common thread that ties together all CAD/CAM developments is the focus on data bases and data flow.

The finite element method has been used in engineering analysis for many years. However, its use throughout the entire engineering design process has been limited by the cost of the computing resources and the workforce needed to synthesize and manage the data for multiple analyses required by design modifications. The drastically reduced expense of computing resources and the development of more integrated design software are increasing the cost-effectiveness of finite element methods to the point that they can become the design process, not just a check on the final design or a tool for postfailure analysis.

The process of extracting the geometry of a design from a computer-aided design database, generating a finite element model, analyzing that model, and retrieving the results for interpretation requires the transmittal of vast amounts of information. Thus, data communication procedures between the various segments are paramount for the effective integration of the process. The two requirements placed on this data transfer are that it be as general, complete, and foolproof as possible and that it be as rapid as possible. There are no methods currently that meet these requirements.

5.3   Building the Basics for the Finite Element Method

The basic concept of the finite element method is not new. It has been used throughout for evaluating certain quantities (particularly area and volume) by adding or counting well-defined geometric figures (elements). Today's understanding of the finite element, however, is finding an approximate solution to a boundary- and initial-value problem by assuming that the domain is divided into well-defined subdomains (elements) and that the unknown function of the state variable is defined approximately within each element. With these individually defined functions matching each other at the element nodes or at the interfaces, the unknown function is approximated over the entire domain.

There are, of course, many other approximate methods for the solution of boundary-value problems, such as finite-difference methods, weighted-residual methods, Rayleigh-Ritz methods, Galerkin methods, and more. The primary difference between the finite element method and most other methods is that in the finite element method the approximation is confined to relatively small subdomains. It is, in a way, the localized version of the Rayleigh-Ritz method. Instead of finding an admissible function satisfying the boundary conditions for the entire domain, which, particularly for irregular domains, is often difficult, in the finite element methods the admissible functions are defined over element domains with simple geometry and pay no attention to complications at the boundaries. This is one of the reasons why the finite element method has gained superiority over the other approximate methods.

Since the entire domain is divided into numerous elements and the function is approximated in terms of its values at certain points (nodes), it is inevitable that the evaluation of such a function will require solution of simultaneous equations. Because of this, the finite element methods were not widely used until the middle of this century when powerful computers were conceived and developed.

5.3.1 Basic steps in the finite element method

Regardless of the physical nature of the problem, a standard finite element method involves the following steps.

Depending upon the physical nature of the problem, the complexity of a solution may or may not require planning the solution in detail.

Step 1. Definition of the Problem and Its Domain. In finite element methods, there are three basic sources of approximation. The first one is the definition of the domain, both physically and geometrically; the second is the discretization of the physical domain; the third is the solution algorithms. The approximations used in defining the physical characteristics of different regions of the domain are very much problem oriented. The geometric definition of the domain, however, requires establishing global coordinate axes in reference to which the coordinates of certain points, which define the equations of the line and surfaces of elements, are to be described. This coordinate system need not be rectangular and cartesian as is often the case; some curvilinear systems may actually be better suited to the specific problem.

The domain can be bounded or unbounded (some portions extend to infinity). For the bounded region of the domain, the idealization is done by using finite elements and for the unbounded portion by using infinite elements or boundary elements. Quite often, the entire domain is made up of subdomains. The interface conditions between subdomains must also be defined prior to discretization.

Step 2. Discretion of the Domain. Since the problem is usually defined over a continuous domain, the governing equations, with the exception of the essential boundary conditions, are valid for the entirety of, as well as for any portion of, that domain. This allows idealization of the domain in the form of interconnected finite-sized domains (elements) of different size and shape. By doing this, certain approximations are introduced. Putting enough numbers of nodes between the elements (higher-order elements, etc.) also comes into the picture at this stage of the method. Here, one should be concerned with how well the idealized discrete domain represents the actual continuous domain. To a certain extent, it is true that the smaller elements (finer mesh) produce better results. But it is also true that the finer mesh results in a larger number of equations to be solved and decreases the accuracy, thereby contrasting with the purpose of using the method. The question then arises: what is the most efficient element type, size, and pattern? A partial answer to this question is given in the literature under the key word modeling. Adaptive processes or mesh refinements and automatic mesh generation are also techniques relevant to the discretization of the domain.

In finite element idealization of the domain, we shall, in general, make reference to the following elements: finite element and master element.

Finite elements are those which, when put together, result in the discrete version of the actual continuous domain. Finite elements are generally straight-sided (or -surfaced), particularly at the interior of the domain. They can be curved (as the higher-order elements) mainly at the boundaries or curved surfaces (as in shells). The curved elements, therefore, may contain geometric approximations in addition to physical approximations. Their geometric approximations are controlled by the number of nodes utilized at the exterior of the elements to define their shape. The physical approximations are controlled by the total number of nodes (exterior as well as interior) utilized in defining the trial functions (shape functions) for the state variable.

Master elements are those which are used in place of finite elements in order to facilitate computations in the element domain. The definition of shape functions and particularly of integration becomes simpler in master elements.

Step 3. Identification of State Variable(s). Until this step, no reference has been made to the physical nature of the problem. Whether it is a heat-transfer problem, fluid- or solid-mechanics problem, etc., comes into the picture at this stage. The mathematical description of steady state physical phenomena, for instance, leads to an elliptic boundary-value problem in which the formula contains the state variable and the flux. These variables are related to each other by a constitutive equation representing a mathematical expression of a physical law.

Once the state variable and the flux have been identified, the formulation can take place containing either or both. The choice is usually dictated by the problem. Various finite element methods, in general, are the result of such choices. Comprehensive discussions of alternative finite element methods in mechanics may be found in Roark, Kardestuncer, and Baker.[23-25]

Step 4. Formulation of the Problem. Very often a physical problem is formulated either via a set of differential equations

$$Lu = f \tag{5.1}$$

with boundary conditions, or by an integral equation (a functional)

$$\pi = \int_{\Omega} G(x, y, z, u)d\Omega + \int_{\Gamma} g(x, y, z, u)d\Gamma \tag{5.2}$$

subject to stationary requirement. While Eq. 5.1 is referred to as the operational form of the physical problem, Eq. 5.2 is referred to as the variational form of the same problem. Solution of either equation (inversion of L or minimization of π) yields the same results.

For the heat conduction problem in two-dimensional space, for instance, Eqs. 5.1 and 5.2 take the following forms:

$$\nabla^2 u - c = 0$$
$$\pi = \int_{A} \left(1/2\, u^2 + cu\right)dA \tag{5.3}$$

In general, if a functional (Eq. 5.2) exists, the associated Euler-Lagrange differential equation (Eq. 5.1) can be found. The reverse, however, is not necessarily true. While the differential equation may be approximated over a set of discrete points using finite differences, the associated functional can be minimized over a set of discrete domains by using the finite element method. A parallel treatment of the two classes of formulation can be found in Strang and Fix.[26]

Step 5. Establishing Coordinate Systems. There are primarily two reasons for choosing special coordinate axes for the elements in addition to the global axes for the entire system. The first is the ease of constructing the trial functions for the elements, and the second is ease of integration within the elements. However, since the elements will be assembled in the global frame, this step introduces additional computations in the form of coordinate transformations. Although the entire finite element analysis can be carried out directly in global system, the benefit does not warrant the price paid for it. Since the coordinate transformations between any two coordinate systems are well defined, using the most suitable coordinate axes for each type of element and for the overall problem is highly recommended.

Depending on the element shape, one usually chooses cartesian or curvilinear axes located within the element in reference to which the element matrix equation will be obtained. Other coordinate systems, known as natural coordinates such as area or volume coordinates, are often employed in finite element analysis because the numerical integration is much simpler with respect to these coordinates.

The transformation of entities from an element to a corresponding master element involves mapping. Transformation from the element coordinate system to the global coordinates, however, involves only rotations. If both coordinate systems, local (x, y, z) and global $(X, Y,$

Z), are chosen so as to be orthogonal, the rotation matrix is also orthogonal. This greatly facilitates the transformation between the two coordinate systems.

Once the coordinate axes are established, the element equations are ordinarily computed first in a master element. They are then finally transformed into the system for assembly. In some cases, however, the mapping of master element computations can be done directly into the global coordinate systems.

After the solution of simultaneous equations, the inverse transformation takes place in order to compute the physical entities required from the element domain.

Step 6. Constructing Approximate Functions for the Elements. Once the state variable(s) and the local coordinate system have been chosen, the function can be approximated in several ways. Only two entities need to be approximated. The first is physical (the state variable) and the second is geometrical (the shape of the element). If the element is actually made up of straight lines or planes, the coordinates of primary nodes (those at the extremes of the elements) will define the element shape accurately. In this case, the geometric approximation does not enter into the picture. Because of this, discretization of the entire domain is usually made by straight-line (linear) elements. For some problems, however, linear elements may introduce unacceptable errors, and discretization has to be accomplished by using isoparametric elements.

A similar argument is valid for the approximation of the state variable. It can be approximated in the form of a linear function or a higher-order function (i.e., quadratic, cubit, etc.). The user must decide whether to approximate physics (state variable) and geometry (shape) equally or to give preference to one or the other in various regions of the domain.

Step 7. Obtain the Element Matrices and Equations. At this point in the game it is assumed that the formulation and discretization of the domain with the desired element shapes and dimensions have been completed. The approximating function for the state variable is written in terms of *shape functions* (derived for each element geometry). This further approximation contains the approximation for the state variable and the coordinates for the element nodes that define the shape of the element. Substitution of this approximation into the variational approach to solving the system (Eq. 5.2) yields, with considerable intensive integration, a system of equations to be solved for the state variables.

Step 8. Coordinate Transformations. The determination of the system of equations for the nodal values involves the integration of the shape functions or their derivatives, or both, over the element. The integration is easier to evaluate when the interpolation equation is written in the terms of the element coordinate system (a coordinate system located on or within the boundaries of the element).

Coordinate transformations of physical entities such as vectors and matrices follow well-defined rules such as those presented in Chap. 4.

Step 9. Assembly of the Element Equations. The assembly of the element matrix equations is done according to the topological configuration of the elements after this equation is transformed into the global system. The assembly is done through the nodes at the interfaces which are common to the adjacent elements. At these nodes the continuities are established with respect to the state variable(s) and also with respect to the derivatives of the state variables.

The essential boundary conditions are introduced so that the final set of equations will be reduced, or perhaps condensed.

Step 10. Solution of the Final Set of Equations. The finite element methods yield the solution of a set of simultaneous equations. The solution procedure can be categorized into three parts: (1) direct, (2) iterative, and (3) stochastic.

The direct-solution techniques (originally proposed by Gauss a century ago) consist of a set of systematic steps, and are used often in finite element solutions. The accuracy of the results is largely determined by the condition of equations (well or ill), the number of equations, and the computer (double precision, single precision, etc.). The symmetry and banded properties of equations can be well taken care of with these methods. The Gauss elimination and Cholesky's factorization (LU decomposition) are the most commonly used direct procedures. These methods are well suited to a small or moderate number of equations.

When systems are of a very large order, iterative procedures such as Gauss-Seidel or Jacobi iterations are more suited. Iterative methods are, in general, self-correcting, and the accuracy of the solution depends upon the number of iterations. Convergence is not always assured, but when convergence does occur, the sequence of equations plays an important role. The solution time is considerably less than that required by a direct procedure. On the other hand, the iterative methods are not suited for sets with multiple right-hand sides (multiple loading conditions in solid-mechanics problems, for example).

When the set of equations to be solved is nonlinear, the Newton-Raphson iteration and its modified version appear to be the most commonly used methods. In recent years, the quasi-Newton method[9] has also received considerable attention in nonlinear finite element analysis. A detailed presentation of the solution of simultaneous equations (linear and nonlinear) by using various procedures and with corresponding algorithms may be found in Kardestuncer.[24]

Step 11. Interpretation of the Results. The previous step resulted in the approximate values of the state variable at discrete points (nodes) of the domain. Normally, these values are interpreted and used for calculation of other physical entities, such as flux either throughout the domain or in certain regions of it.

We have elected to present a finite element solution to the two-dimensional heat transfer equations typical of that presented in Chap. 8.

Typical examples of this type of problem include steady seepage through soils and steady heat flow through a conductor. The program given is for two-dimensional planar conditions and uses four node quadrilateral elements. Each node has only one unknown or degree of freedom associated with it. The unknown would represent, in the case of seepage problems, the fluid potential and, in the case of conduction problems, the temperature.

Systems that are governed by LaPlace's equation require boundary conditions to be prescribed at all points around a closed domain. These boundary conditions commonly take the form of fixed values of the potential or values of the first derivative of the potential normal to the boundary. The problem amounts to finding the values of the fluid potential at points within the closed domain.

Being elliptic in character, solution of LaPlace's equation quite closely resembles the solution of the equilibrium equations in solid elasticity. Both methods ultimately require the solution of a set of linear simultaneous linear equations. The element stiffness matrices are formed numerically and assembled into a global stiffness matrix which is symmetrical and banded.

It is not the purpose of this text to go into detail behind the subroutines presented in the following sections. There are many texts that consider this subject.

5.3.2   General flow diagram for solving sample problem

The following unstructured flow diagram is presented to illustrate the flow of the program.

Reserve space for all fixed and variable dimension arrays.

Read data and initialize arrays.

For All Elements:

Find nodal coordinates.

Determine the steering vector.

Null the element matrices.

Find the shape functions.

Convert to global coordinates.

Form the stiffness matrix.

Add contributions from boundary conditions.

Form global stiffness matrix.

Solve equations.

Print results.

5.3.3   The problem

The problem under consideration consists of a rectangular area with dimensions of 6 × 4. The top surface is held at a temperature of 0°F and the right side is fixed at a temperature of 50°F. The slab is insulated at the other two faces. Material properties for aluminum are used for this two-dimensional planar problem. The problem is illustrated in Fig. 5.1. The numbers shown at the corners are the node numbers and the numbers within the elements are the element numbers. Also shown are the top and right edge temperatures.

Problem specifics are listed below:

CONTROL INFORMATION

number of node points	=	35
number of element types	=	2
number of load cases	=	1
gravitational constant	=	3.8640E+02

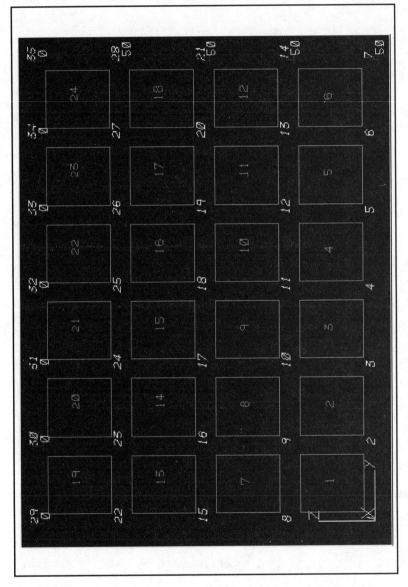

Figure 5.1 Model elements with node and element numbers.

NODAL DATA

| NODE | BOUNDARY CONDITION CODES | | | | | | NODAL POINT COORDINATES | | | |

NO.	DX	DY	DZ	RX	RY	RZ	X	Y	Z	T
1	1	0	1	1	1	1	0.000E+00	0.000E+00	0.000E+00	0.000E+00
2	1	0	1	1	1	1	0.000E+00	1.000E+00	0.000E+00	0.000E+00
3	1	0	1	1	1	1	0.000E+00	2.000E+00	0.000E+00	0.000E+00
4	1	0	1	1	1	1	0.000E+00	3.000E+00	0.000E+00	0.000E+00
5	1	0	1	1	1	1	0.000E+00	4.000E+00	0.000E+00	0.000E+00
6	1	0	1	1	1	1	0.000E+00	5.000E+00	0.000E+00	0.000E+00
7	1	0	1	1	1	1	0.000E+00	6.000E+00	0.000E+00	0.000E+00
8	1	0	1	1	1	1	0.000E+00	0.000E+00	1.000E+00	0.000E+00
9	1	0	1	1	1	1	0.000E+00	1.000E+00	1.000E+00	0.000E+00
10	1	0	1	1	1	1	0.000E+00	2.000E+00	1.000E+00	0.000E+00
11	1	0	1	1	1	1	0.000E+00	3.000E+00	1.000E+00	0.000E+00
12	1	0	1	1	1	1	0.000E+00	4.000E+00	1.000E+00	0.000E+00
13	1	0	1	1	1	1	0.000E+00	5.000E+00	1.000E+00	0.000E+00
14	1	0	1	1	1	1	0.000E+00	6.000E+00	1.000E+00	0.000E+00
15	1	0	1	1	1	1	0.000E+00	0.000E+00	2.000E+00	0.000E+00
16	1	0	1	1	1	1	0.000E+00	1.000E+00	2.000E+00	0.000E+00
17	1	0	1	1	1	1	0.000E+00	2.000E+00	2.000E+00	0.000E+00
18	1	0	1	1	1	1	0.000E+00	3.000E+00	2.000E+00	0.000E+00
19	1	0	1	1	1	1	0.000E+00	4.000E+00	2.000E+00	0.000E+00
20	1	0	1	1	1	1	0.000E+00	5.000E+00	2.000E+00	0.000E+00

21	1	0	1	1	1	1	0.000E+00	6.000E+00	2.000E+00	0.000E+00
22	1	0	1	1	1	1	0.000E+00	0.000E+00	3.000E+00	0.000E+00
23	1	0	1	1	1	1	0.000E+00	1.000E+00	3.000E+00	0.000E+00
24	1	0	1	1	1	1	0.000E+00	2.000E+00	3.000E+00	0.000E+00
25	1	0	1	1	1	1	0.000E+00	3.000E+00	3.000E+00	0.000E+00
26	1	0	1	1	1	1	0.000E+00	4.000E+00	3.000E+00	0.000E+00
27	1	0	1	1	1	1	0.000E+00	5.000E+00	3.000E+00	0.000E+00
28	1	0	1	1	1	1	0.000E+00	6.000E+00	3.000E+00	0.000E+00
29	1	0	1	1	1	1	0.000E+00	0.000E+00	4.000E+00	0.000E+00
30	1	0	1	1	1	1	0.000E+00	1.000E+00	4.000E+00	0.000E+00
31	1	0	1	1	1	1	0.000E+00	2.000E+00	4.000E+00	0.000E+00
32	1	0	1	1	1	1	0.000E+00	3.000E+00	4.000E+00	0.000E+00
33	1	0	1	1	1	1	0.000E+00	4.000E+00	4.000E+00	0.000E+00
34	1	0	1	1	1	1	0.000E+00	5.000E+00	4.000E+00	0.000E+00
35	1	0	1	1	1	1	0.000E+00	6.000E+00	4.000E+00	0.000E+00

number of elements   =   24

number of materials   =   1

maximum temperatures per material   =   1

MATERIAL PROPERTIES

weight density   =   9.8000E-02

mass   density   =   2.5360E-04

Stefan-Boltzmann constant   =   3.3025E-15

absolute temperature conversion =   4.5970E+02

TEMPERATURE	K(N)	K(S)	K(T)	SP. HEAT
.0	2.315E-03	2.315E-03	2.315E-03	2.140E-01

ELEMENT CONNECTIVITY DATA

ELEM NO.	NODE I	NODE J	NODE K	NODE L
1	1	2	9	8
2	2	3	10	9
3	3	4	11	10
4	4	5	12	11
5	5	6	13	12
6	6	7	14	13
7	8	9	16	15
8	9	10	17	16
9	10	11	18	17
10	11	12	19	18
11	12	13	20	19
12	13	14	21	20
13	15	16	23	22
14	16	17	24	23
15	17	18	25	24
16	18	19	26	25

17	19	20	27	26
18	20	21	28	27
19	22	23	30	29
20	23	24	31	30
21	24	25	32	31
22	25	26	33	32
23	26	27	34	33
24	27	28	35	34

TEMPERATURE ELEMENTS

number of elements =   11

ELEMENT CONNECTIVITY DATA

ELEMENT NUMBER	NODE N	CODE KD	SPECIFIED TEMPERATURE	THERMAL STIFFNESS
1	7	1	5.0000E+01	1.0000E+01
2	14	1	5.0000E+01	1.0000E+01
3	21	1	5.0000E+01	1.0000E+01
4	28	1	5.0000E+01	1.0000E+01
5	29	1	0.0000E+00	1.0000E+01
6	30	1	0.0000E+00	1.0000E+01
7	31	1	0.0000E+00	1.0000E+01
8	32	1	0.0000E+00	1.0000E+01

9	33	1	0.0000E+00	1.0000E+01
10	34	1	0.0000E+00	1.0000E+01
11	35	1	0.0000E+00	1.0000E+01

5.3.4   Variable definitions for sample FEA program

Before giving the program solution to the problem, it is worthwhile to list the variables used in the main program and subroutines with a short description of the variable.

NEX	number of elements in x direction
NEY	number of elements in y direction
N	number of degrees of freedom in mesh
IW	half-bandwidth of mesh
NN	number of nodes in mesh
NR	number of restrained nodes in mesh
NGP	number of integrating points per dimension
AA	dimension of elements in x direction
BB	dimension of elements in y direction
CONDX	conduction in x direction
CONDY	conduction in y direction
IFIX	number of prescribed freedoms
IDER	size of shape function (DER) matrix (local)
IDERIV	size of shape function (DERIV) matrix (global)
IJAC	size of jacobian matrix
IJACI	size of inverse jacobian matrix
ICOND	size of conductivity matrix
IKDERV	size of product matrix COND*DERIV
IWGTSAMP	size of array for Gauss quadrature weights
IDTKD	size of prod. array DERIVT*COND*DERIV
IKP	size of element stiffness matrix
ICOORD	size of element coordinate matrix
IDERVT	size of DERIV transpose
IKV	size of global stiffness matrix

ILOADS	size of global potentials
INF	size of nodal freedom array
INO	size of prescribed freedom array
IR	N*(IW + 1) – working length of vectors KV and KVH
IT	dimensions of problem
NOD	number of nodes per element
NODOF	number of freedoms per node
DET	determinant of element jacobian matrix
QUOT	scaled weighting coefficient
TEMP	nodal temperatures
WGTSAMP	quadrature abscissae and weights
JAC	jacobian matrix
JACI	inverse of jacobian matrix
CON	conduction matrix
DTKD	product of DERIVT*CON*DERIV
KP	element stiffness matrix
COORD	element nodal coordinates
DER	derivatives of shape functions in local coordinates
DERIV	derivatives of shape functions in global coordinates
DERIVT	transpose of DERIV
KDERIV	product CON*DERIV
FUNCTION	element shape functions in local coordinates
G	element steering vector
KV	global stiffness matrix (IKV > IR)
KVH	copy of KV
LOADS	global potentials (ILOADS > N)
DISPS	calculated temperatures
NF	nodal freedom array (TNF > NN)
NO	prescribed freedoms (TNO > IFTX)
VAL	value to be fixed

5.4   The Finite Element Program

The following is a listing of a FORTRAN FEA program for solving a 2-dimensional heat conduction problem.

```
C     SOLUTION OF TWO-DIMENSIONAL HEAT CONDUCTION EQUATION OVER
C     A PLANE AREA USING 4-NODE QUADRILATERALS
C
      PARAMETER(IKV=1000,ILOADS=150,INF=50,INO=10)
C
      REAL JAC(2,2),JACI(2,2),CON(2,2),WGTSAMP(3,2),DTKD(4,4),KP(4,4),
     1COORD(4,2),DER(2,4),DERIV(2,4),DERIVT(4,2),KDERIV(2,4),FUNCTION(4),
     1VAL(INO),KVH(IKV),KV(IKV),LOADS(ILOADS),DISPS(ILOADS)
      INTEGER G(4),NO(INO),NF(INF,1)
      DATA IT,IJAC,IJACI,ICON,IDER,IDERIV,IKDERV/7*2/,IWGTSAMP/3/
      DATA IDTKD,IKP,ICOORD,IDERVT,NOD/5*4/,NODOF/1/
C
C     INPUT AND INITIALIZATION
C
      READ(5,*) NEX,NEY,N,IW,NN,NR,NGP,AA,BB,CONDX,CONDY
      CALL READNF(NF,INF,NN,NODOF,NR)
      IR=N*(IW+1)
      CALL NULVEC(KV,IR)
      CALL NULL(CON,ICON,IT,IT)
      CON(1,1)=CONDX
      CON(2,2)=CONDY
      CALL GAUSLEGN(WGTSAMP,IWGTSAMP,NGP)
C
C     ELEMENT INTEGRATION AND ASSEMBLY
C
      DO 10 IP=1,NEX
      DO 10 IQ=1,NEY
      CALL QUADGEO(IP,IQ,NEX,AA,BB,COORD,ICOORD,G,NF,INF)
      CALL NULL(KP,IKP,NOD,NOD)
      DO 20 I=1,NGP
      DO 20 J=1,NGP
      CALL SHAPE(DER,IDER,FUNCTION,WGTSAMP,IWGTSAMP,I,J)
      CALL MATMUL(DER,IDER,COORD,ICOORD,JAC,IJAC,IT,NOD,IT)
      CALL INV2X2(JAC,IJAC,JACI,IJACI,DET)
      CALL MATMUL(JAC1,IJAC1,DER,IDER,DERIV,IDERIV,IT,IT,NOD)
      CALL MATMUL(CON,ICON,DERIV,IDERIV,KDERIV,IKDERV,IT,IT,NOD)
      CALL MATRAN(DERIVT,IDERVT,DERIV,IDERIV,IT,NOD)
      CALL MATMUL(ERIVT,IDERVT,KDERIV,IKDERV,DTKD,IDTKD,NOD,IT,NOD)
      QUOT=DET*WGTSAMP(I,2)*WGTSAMP(J,2)
      CALL MSMULT(DTKD,IDTKD,QUOT,NOD,NOD)
   20 CALL MATADD(KP,IKP,DTKD,IDTKD,NOD,NOD)
   10 CALL GBLSTIF(KV,KP,IKP,G,N,NOD)
      CALL VECCOPY(KV,KVH,IR)
C
C     SPECIFY FIXED POTENTIALS AND REDUCE EQUATIONS
C
      READ(5,*)IFIX,(NO(I),VAL(I),I.I,IFIX)
```

```
        CALL NULVEC(LOADS,N)
        DO 30 I=1,IFIX
        KV(NO(I))=KV(NO(I))+1.E20
 30     LOADS(NO(I))=KV(NO(I))*VAL(I)
        CALL BWREDUC(KV,N,IW)
C
C       SOLVE EQUATIONS AND RETRIEVE TEMPERATURES
C
        CALL BACSUB(KV,LOADS,N,IW)
        CALL PRINTV(LOADS,N)
        CALL MATVECMU(KVH,LOADS,DISPS,N,IW)
        TEMP=0.
        DO 40 I=1,IFIX
 40     TEMP=DISPS(NO(I))
        WRITE(6,'(E12.4)')TEMP
        STOP
        END

        SUBROUTINE BACSUB(BK,LOADS,N,IW)
C
C       THIS SUBROUTINE PERFORMS THE GAUSSIAN BACK-SUBSTITUTION
C
        REAL BK(*),LOADS(*)
        LOADS(I).LOADS(I)/BK(I)
        DO 1 I=2,N
        SUM=LOADS(I)
        I1=I-1
        NKB=I-IW
        IF(NKB)2,2,3
 2      NKB=I
 3      DO 4 K=NKB,I1
        JN=(I-K)*N+K
        SUM=SUM-BK(JN)*LOADS(K)
 4      CONTINUE
        LOADS(I)=SUM/BK(I)
 1      CONTINUE
        DO 5 JJ=2,N
        I=N-JJ+1
        SUM=0.
        I1=I+1
        NKB=I+IW
        IF(NKB-N)7,7,6
 6      NKB=N
 7      DO 8 K=I1,NKB
        JN=(K-I)*N+I
 8      SUM=SUM+BK(JN)*LOADS(K)
        LOADS(I)=LOADS(I)-SUM/BK(I)
 5      CONTINUE
        RETURN
        END
```

```
      SUBROUTINE BWREDUC(BK,N,IW)
C
C     THIS SUBROUTINE PERFORMS GAUSSIAN REDUCTION OF
C     THE STIFFNESS MATRIX STORED AS A VECTOR BK(N*(IW+1))
C
      REAL BK(*)
      DO 1 I=2,N
      IL1=I-1
      KBL=IL1+IW+I
      IF(KBL-N)3,3,2
    2 KBL.N
    3 DO 1 J=I,KBL
      LJ=(J-I)*N+I
      SUM=BK(LJ)
      NKB=J-IW
      IF(NKB)4,4,5
    4 NKB=I
    5 IF(NKB-IL1)6,6,8
    6 DO 7 M=NKB,IL1
      NI=(I-M)*N+M
      NJ=(J-M)*N+M
    7 SUM=SUM-BK(NI)*BK(NJ)/BK(M)
    8 BK(LJ).SUM
    1 CONTINUE
      RETURN
      END


      SUBROUTINE GBLSTIF(BK,KM,IKM,G,N,IDOF)
C     THIS SUBROUTINE FORMS THE GLOBAL STIFFNESS MATRIX
C     STORING THE UPPER TRIANGLE AS A VECTOR BK(N*(IW+1))
C
      REAL BK(*),KM(IKM,*)
      INTEGER G(*)
      DO 1 I=1,IDOF
      IF(G(I).EQ.O)GOTO 1
      DO 5 J=1,IDOF
      IF(G(J).EQ.0)GOTO 5
      ICD=G(J)-G(I)+1
      IF(ICD-1)5,4,4
    4 IVAL=N*(ICD-1)+G(I)
      BK(IVAL)=BK(IVAL)+KM(I,J)
    5 CONTINUE
    1 CONTINUE
      RETURN
      END


      SUBROUTINE SHAPE(DER,IDER,FUNCTION,WGTSAMP,IWGTSAMP,I,J)
C
C     THIS SUBROUTINE FORMS THE SHAPE FUNCTIONS AND
C     THEIR DERIVATIVES FOR 4-NODED QUAD ELEMENT
C
```

```
      REAL DER(IDER,*),FUNCTION(*),WGTSAMP(IWGTSAMP,*)
      ETA=WGTSAMP(I,I)
      XI=WGTSAMP(J,l)
      ETAM=.25*(l.-ETA)
      ETAP=.25*(1.+ETA)
      XIM=.25*(1.-XI)
      XIP=.25*(1.+XI)
      FUNCTION(1)=4.*XIM*ETAM
      FUNCTION(2)=4-*XIM*ETAP
      FUNCTION(3)=4.*XIP*ETAP
      FUNCTION(4)=4.*XIP*ETAM
      DER(l,l)=-ETAM
      DER(l,2)=-ETAP
      DER(l,3)=ETAP
      DER(l,4)=ETAM
      DER(2,l)=-XIM
      DER(2,2)=XIM
      DER(2,3)=XIP
      DER(2,4)=-XIP
      RETURN
      END


      SUBROUTINE MATVECMU(BK,DIS PS,LOADS,N,IW)
C     THIS SUBROUTINE MULTIPLIES A MATRIX BY A VECTOR
C     THE MATRIX IS SYMMETRICAL WITH ITS UPPER TRIANGLE
C     STORED AS A VECTOR
      REAL BK(*),DISPS(*),LOADS(*)
      DO 1 I=1,N
      X=0.
      DO 2 J=1,IW+1
      IF(I+J.LE.N+1)X=X+BK(N*(J 1)+I)*DISPS(I+J- 1)
   2  CONTINUE
      DO 3 J=2,IW+1
      IF(I-J +1.GE.1)X=X+BK((N-1)*(J-1)+I)* DISPS(I-J+1)
   3  CONTINUE
      LOADS(I)=X
   1  CONTINUE
      RETURN
      END


      SUBROUTINE MATMUL(A,IA,B,IB,C,IC,L,M,N)
C
C     THIS SUBROUTINE FORMS THE PRODUCT OF TWO MATRICES
C
      REAL A(IA,*),B(IB,*),C(IC,*)
      DO 1 I=1,L
      DO 1 J=1,N
      X=0.0
      DO 2 K.I,M
   2  X=X+A(I,K)*B(K,J)
```

```
      C(I,J)=X
   1    CONTINUE
      RETURN
      END


      SUBROUTINE MATRAN(A,IA,B,IB,M,N)
C
C       THIS SUBROUTINE FORMS THE TRANSPOSE OF A MATRIX
C
      REAL A(IA,*),B(IB,*)
      DO 1 I=1,M
      DO 1 J=1,N
   1    A(J,I)=B(I,J)
      RETURN
      END


      SUBROUTINE MVMULT(M,IM,V,K,I,Y)
C
C       THIS SUBROUTINE MULTIPLIES A MATRTX BY A VECTOR
C
      REAL M(IM,*),V(*),Y(*)
      DO 1 I=1,K
      X=0.
      DO 2 J=1,L
   2    X=X+M(I,J)*V(J)
      Y(I)=X
   1    CONTINUE
      RETURN
      END


      SURROUTINE NULL(A,IA,M,N)
C
C       THIS SUBROUTINE NULLS A 2-D ARRAY
C
      REAL A(IA,*)
      DO 1 I=1,M
      DO 1 J=1,N
   1    A(I,J)=0.0
      RETURN
      END


      SUBROUTINE NULVEC(VEC,N)
C
C       THIS SUBROUTINE NULLS A COLUMN VECTOR
C
      REAL VEC(*)
      DO 1 I=1,N
   1    VEC(I).0.
      RETURN
      END
```

```
      SUBROUTINE MATADD(A,IA,B,IB,M,N)
C
C     THIS SUBROUTINE ADDS TWO EQUAL SIZED ARRAYS
C
      REAL A(IA,*),B(IB,*)
      DO 1 I=1,M
      DO 1 J=1,N
   1  A(I,J)=A(I,J)+B(I,J)
      RETURN
      END

      SUBROUTINE INV2X2(JAC,IJ AC,JACI,IJACI,DET)
C
C     THIS SUBROUTINE FORMS THE INVERSE OF A 2 BY 2 MATRIX
C
      REAL JAC(IJAC,*),JACI(IJACI,*)
      DET=JAC(1,1)*JAC(2,2)-JAC(1,2)*JAC(2,1)
      JACI(1,1)=JAC(2,2)
      JACI(1,2)=-JAC(1,2)
      JACI(2,1)=-JAC(2,1)
      JACI(2,2)=JAC(1,1)
      DO 1 K=1,2
      DO 1 L=1,2
   1  JACI(K,L)=JACI(K,L)/DET
      RETURN
      END


      SUBROUTINE VECCOPY(A,B,N)
C
C     THIS SUBROUTINE COPIES VECTOR A INTO VECTOR B
C
      REAL A(*),B(*)
      DO 1 I=1,N
   1  B(I)=A(I)
      RETURN
      END

      SUBROUTINE GAUSLEGN(WGTSAMP,IWGTSAMP,NGP)
C
C     THIS SUBROUTINE PROVIDES THE WEIGHTS AND SAMPLING POINTS
C     FOR GAUSS-LEGENDRE QUADRATURE
C
      REAL WGTSAMP(IWGTSAMP,*)
      GO T0(1,2,3,4,5,6,7),NGP
   1  WGTSAMP(1,1)=0
      WGTSAMP(1,2)=2.
      GOTO 100
   2  WGTSAMP(1,1)=1./SQRT(3.)
      WGTSAMP(2,1).-WGTSAMP(I,I)
```

```
        WGTSAMP(l,2)=1.
        WGTSAMP(2,2)=1.
        GO TO 100
3       WGTSAMP(l,l)=.2*SQRT(l5.)
        WGTSAMP(2,l)=.0
        WGTSAMP(3,l)=-WGTSAMP(1,1)
        WGTSAMP(l,2)=5./9.
        WGTSAMP(2,2)=8./9.
        WGTSAMP(3,2)=WGTSAMP(1,2)
        GO TO 100
4       WGTSAMP(l,l)=.861136311594053
        WGTSAMP(2,1)=.339981043584856
        WGTSAMP(3,l)=-WGTSAMP(2,l)
        WGTSAMP(4,l)=-WGTSAMP(l,l)
        WGTSAMP(l,2)=.347854845137454
        WGTSAMP(2,2)=.652145154862546
        WGTSAMP(3,2)=WGTSAMP(2,2)
        WGTSAMP(4,2)=WGTSAMP(l,2)
        GO TO 100
5       WGTSAMP(l,l)=.906179845938664
        WGTSAMP(2,l)=.538469310105683
        WGTSAMP(3,l)=.0
        WGTSAMP(4,l)=-WGTSAMP(2,l)
        WGTSAMP(5,l)=-WGTSAMP(l,l)
        WGTSAMP(l,2)=.236926885056189
        WGTSAMP(2,2)=.478628670499366
        WGTSAMP(3,2)=.568888888888889
        WGTSAMP(4,2)=WGTSAMP(2,2)
        WGTSAMP(5,2)=WGTSAMP(l,2)
        GO TO 100
6       WGTSAMP(l,l)=.932469514203152
        WGTSAMP(2,1)=.661209386466265
        WGTSAMP(3,l)=.238619186083197
        WGTSAMP(4,l)=-WGTSAMP(3,l)
        WGTSAMP(5,l)=-WGTSAMP(2,l)
        WGTSAMP(6,l)=-WGTSAMP(l,l)
        WGTSAMP(l,2)=.171324492379170
        WGTSAMP(2,2)=.360761573048139
        WGTSAMP(3,2)=.467913934572691
        WGTSAMP(4,2)=WGTSAMP(3,2)
        WGTSAMP(5,2)=WGTSAMP(2,2)
        WGTSAMP(6,2)=WGTSAMP(l,2)
        GO TO 100
7       WGTSAMP(l,l)=.949107912342759
        WGTSAMP(2,l)=.741531185599394
        WGTSAMP(3,l)=.405845151377397
        WGTSAMP(4,l)=.0
        WGTSAMP(5,l)=-WGTSAMP(3,l)
        WGTSAMP(6,l)=-WGTSAMP(2,l)
        WGTSAMP(7,l)=-WGTSAMP(l,l)
```

```
        WGTSAMP(l,2)=.129484966168870
        WGTSAMP(2,2)=.279705391489277
        WGTSAMP(3,2)=.381830050505119
        WGTSAMP(4,2)=.417959183473469
        WGTSAMP(5,2)=WGTSAMP(3,2)
        WGTSAMP(6,2)=WGTSAMP(2,2)
        WGTSAMP(7,2)=WGTSAMP(l,2)
  100   CONTINUE
        RETURN
        END


        SUBROUTINE QUADGEO(IP,IQ,NEX,AA,BB,COORD,ICOORD,G,NF,INF)
C
C       THIS SUBROUTINE FORMS THE COORDINATES AND STEERING VECTOR
C       FOR 4-NODE QUADS COUNTING IN X-DIRECTION
C       LAPLACE'S EQUATION 1-FREEDOM PER NODE
C
        REAL COORD(ICOORD,*)
        INTEGER NUM(4),G(*),NF(INF,*)
        NUM(l)=IQ*(NEX+1)+IP
        NUM(2)=(IQ-l)*(NEX+1)+IP
        NUM(3)=NUM(2)+1
        NUM(4)=NUM(l)+1
        DO 1 I=1,4
    1   G(I).NF(NUM(I),I)
        COORD(l,l)=(IP-l)*AA
        COORD(2,l)=(IP-l)*AA
        COORD(3,l)=IP*AA
        COORD(4,l)=IP*AA
        COORD(l,2)=-IQ*BB
        C00RD(2,2)=-(IQ-l)*BB
        C00RD(3,2)=-(IQ-l)*BB
        C00RD(4,2)=-IQ*BB
        RETURN
        END
```

5.5   Solution Results

The temperature results are listed below and illustrated in
Fig. 5.2.

Temperatures of Nodes

Node number	Temperature
1	1.1442D+01
2	1.2348D+01
3	1.5188D+01
4	2.0297D+01

(Continued)

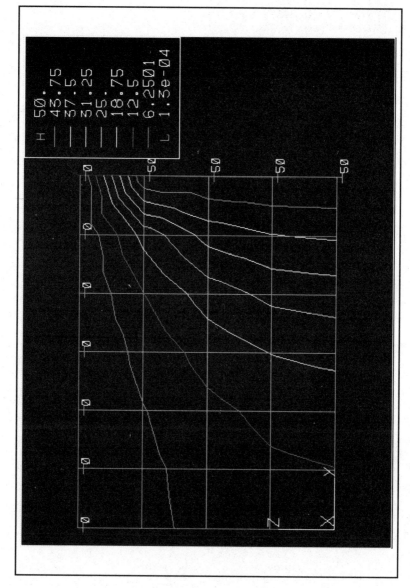

Figure 5.2 Results of analysis.

Node number	Temperature
5	2.8004D+01
6	3.8209D+01
7	5.0000D+01
8	1.0577D+01
9	1.1421D+01
10	1.4079D+01
11	1.8928D+01
12	2.6587D+01
13	3.7248D+01
14	5.0000D+01
15	8.1071D+00
16	8.7652D+00
17	1.0862D+01
18	1.4821D+01
19	2.1590D+01
20	3.3596D+01
21	5.0000D+01
22	4.3939D+00
23	4.7568D+00
24	5.9255D+00
25	8.1991D+00
26	1.2394D+01
27	2.0947D+01
28	4.9999D+01
29	5.3659D−05
30	1.1634D−04
31	1.4570D−04
32	2.0463D−04
33	3.2055D−04
34	6.4310D−04
35	3.5455D−04

FEA in Tools, Molds, and Dies

6.1 Background and Introduction

Beginning in this chapter and continuing to the remainder of the text are the applications of FEA to problems encountered in the manufacturing environment. Many of the problems come from actual applications in various industries. These applications are case studies of obtaining solutions to unique problems. It is the purpose of these problems to show how to use FEA from model building to obtaining a solution for the problems. The first set of problems comes from Moore Special Tool Co., Inc. of Bridgeport, Connecticut.[27]

Moore Special Tool Co., Inc. (Moore) of Bridgeport, Connecticut is an internationally-known manufacturer of high precision Jig Grinders, Aspheric Generators, Universal Coordinate Measuring Machines, and Metrology products.

After Moore began using the popular CADKEY three-dimensional design and drafting software for some of their engineering work, they looked for other computer-aided engineering tools, most notably finite element analysis (FEA), software. "While we were looking for FEA software, one of our major concerns was how the FEA programs would interact with the CAD software that we had already invested in," recalls Harold Lawson of Moore. "We needed a system that would be CADKEY-compatible. ALGOR's FEA System has the ability to convert CADKEY files directly through their CAD program, SuperDraw II, through use of the File Import and Export commands." Equally important for Moore in their decision to use ALGOR's FEA System was the fact that PC-based computer analysis is much faster than the physical testing that Moore had been using.

Before Moore began using ALGOR's finite element analysis, the method of testing a design was to go through the whole process of fabricating the part and trying it out—but that still didn't reveal how manufacturing tolerances would affect it. That is, Moore wouldn't know what would happen if they ran plus or minus 10 or 15 percent off the actual specifications unless they were to go through the trouble of making parts that were 10 to 15 percent off the actual specifications and test them, too. Unfortunately, that required the use of a lot of personnel and manufacturing time. Application of FEA at Moore has meant a substantial time savings in their prototype cycle by enabling them to come much closer to the optimal design the first time.

Since Moore has very close tolerances, prototype testing created a real bottleneck in the design process. Most manufacturers are not worried about millionths of an inch; however, Moore is very concerned.

6.2 The Problem

The first problem presented herein is a part that Moore has analyzed with ALGOR. It is a classic shrink-fit problem involving a plain, thin ring gear having an inner diameter of approximately 6 inches and measuring 13/16 inch high with a cross section of 0.180 inch. This gear is shrink-fitted one-third of the way down from the top of a thick, slightly cone-shaped cylinder. The contact pressure between the inside of the gear and the outside of the cylinder has to be high enough to maintain a friction torque; but if the interference fit is too high, the gear and cylinder will deform. Moreover, the higher the interference fit, the more the cylinder inner diameter will deform. Moore's design goal was to minimize the deformation of the inside cylinder while still producing the required torque transmission between the cylinder and gear. Moore also wanted to know the tolerances and how they would affect the torque transmission capacity of the gear.

Moore could have looked up an approximate shrink fit in the *ASME Handbook* and used their rules. But there was a problem: in the case when a thin ring gear is being shrink-fitted onto a thick, hollow shaft, the *Handbook* does not provide a way to determine the minimum amount of interference fit required; there really isn't one that can be applied specifically to this case. Moore could have attempted to figure it out using the torque and contact pressure of the two surfaces, but this still left yet another question: how much would the inside of the thick cylinder deform due to the pressure from the ring gear? The answer to a question like this is critical to Moore; is it 20 or 40 millionths of an inch?

Furthermore, how much material has to be left so that the toolmakers (who hand-fit and hand-assemble these parts) can do an accurate job? Moore uses CNC grinders, lathes, and milling machines to manufacture their parts; but for hand-fitted parts in general, a hand-lapping and hand-scraping process is employed to remove material from two adjoining surfaces to create a perfect match. The materials from which the gear and cylinder will be fabricated are two types of steel: 4140 and 4150. Both these substances withstand Moore's heat-treating very well.

6.3 The Finite Element Approach to Solving the Problem

The ALGOR FEA System was an indispensable tool that Moore used to design the shrink-fitted gear. The outline geometry of the model is shown in Fig. 6.1. There are two circles of 5 and 6.309998 inches, respectively. These entities are meshed into 16 sectors, each sector representing 22.5° of the model. Although it is only necessary to model a pie-section of the gear-cylinder interface due to its symmetry about the center axis and the uniformity of the loading conditions, it was decided to model the full 360° for documentation and display purposes. To create the full model, the plan view shown in Fig. 6.1 is extruded in the z-direction using the Modify:Copy command and is shown in Fig. 6.2. Eleven increments of 0.81 inch were used to obtain the full height of 8.91 inches.

Boundary conditions on the model are set with the Add:FEA add command and are shown in Figs. 6.3 and 6.4. Several nodes are constrained in the direction of the Z-axis to hold the gear fixed in space.

Completion of the model leads to the transfer of the graphics information presented in SuperDraw II to the form required for the processor. This task is accomplished with the *Transfer* command as given in Fig. 6.5. At this stage of the analysis, all information concerning the elements and their properties are added to the graphical information. In addition, it is at this point that a uniform internal pressure of 284 psi is applied to each of the element faces on the outer diameter of the cylinder. The value of 284 psi for the pressure on the cylinder face and gear ring interface was estimated from the values in the *ASME Handbook*. This value allows a safety factor of four (4) in the design to be able to transmit 90 ft·lb of torque. Figure 6.6 illustrates the decoder screen.

After the decoder, the stress menu given in Fig. 6.7 is selected. Figure 6.8 shows the model with arrows representing the pressure on the face of the cylinder. This figure is from the view menu. The next step is to select the processor, in this case the Stress Processor, for the analysis.

Figures 6.9 and 6.10 illustrate the results of the analysis of the cylinder. Given the material, deflections and stress levels are well within requirements.

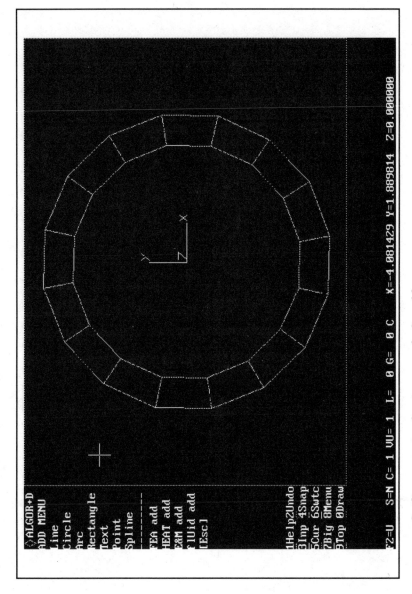

Figure 6.1 Outline geometry for ring gear shaft problem.

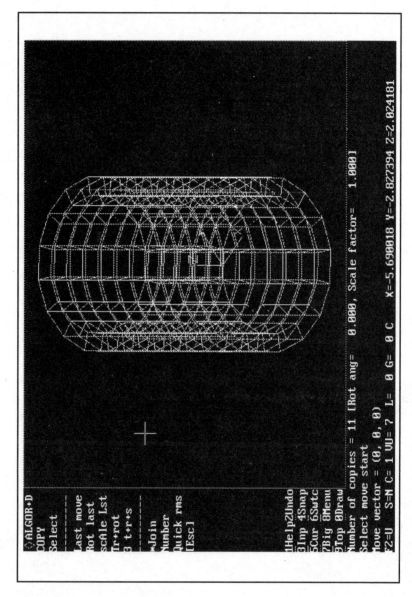

Figure 6.2 Extruded shaft for ring gear problem.

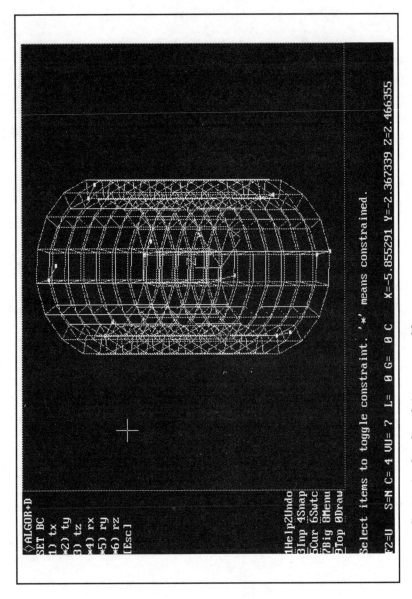

Figure 6.3 Constrained nodes of ring gear problem.

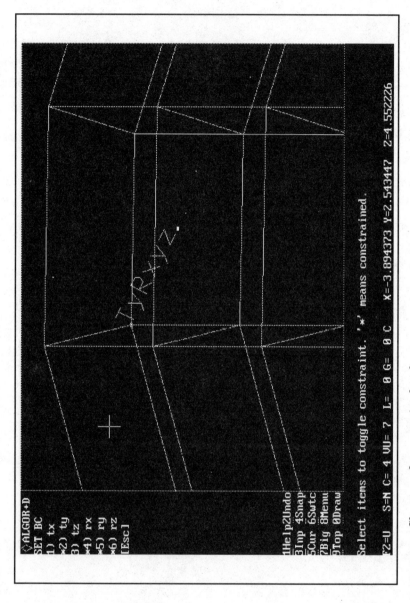

Figure 6.4 Close-up of a constrained node.

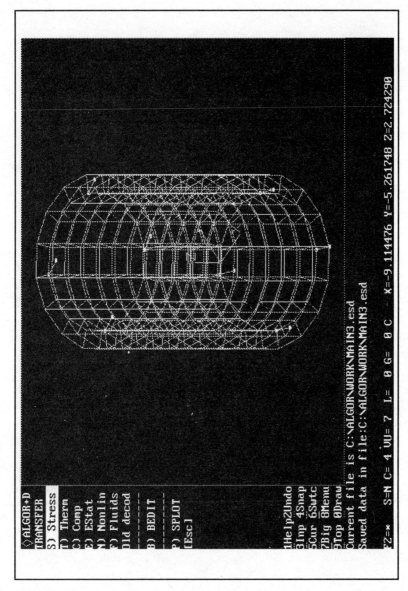

Figure 6.5 Transfer of graphical model to preprocessor.

137

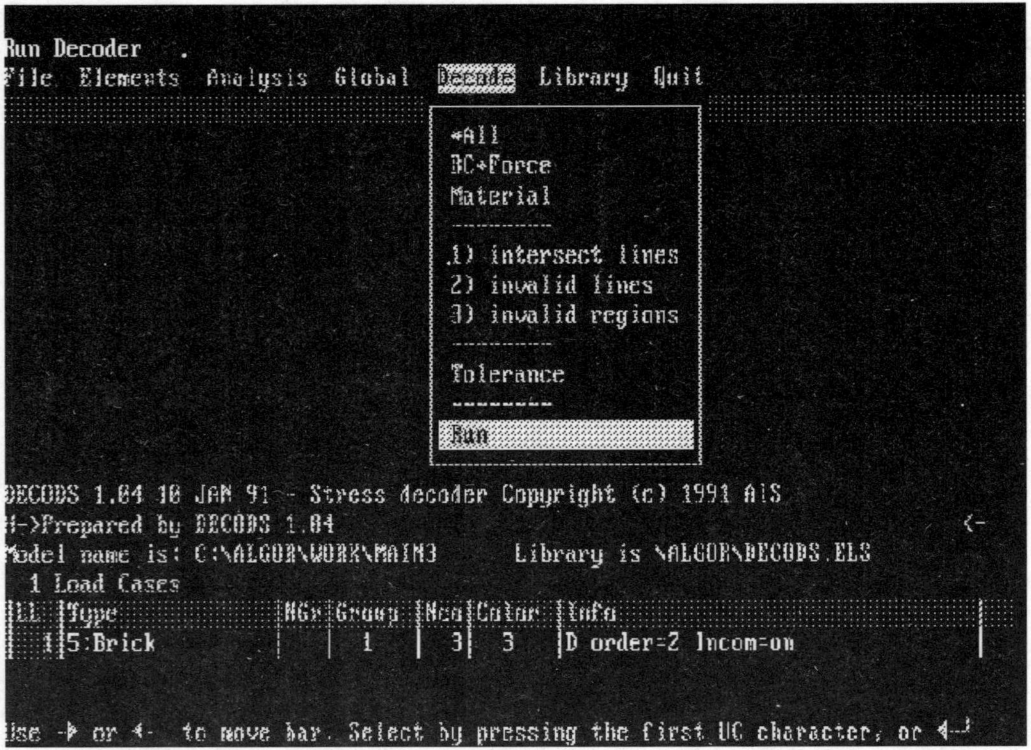

Run Decoder .
File Elements Analysis Global Decode Library Quit

*All
BC+Force
Material

1) intersect lines
2) invalid lines
3) invalid regions

Tolerance

Run

DECODS 1.04 10 JAN 91 - Stress Decoder Copyright (C) 1991 AIS
#->Prepared by DECODS 1.04
Model name is: C:\ALGOR\WORK\WA1M3   Library is \ALGOR\DECODS.ELS
1 Load Cases

EL Type   BC: G-gen Neg/Color Info
115:Brick | 1 | 3 | 3 | 1D order=2 Incom=on

se -> or <- to move bar. Select by pressing the first UC character, or <-|

Figure 6.6   Decoder screen for preprocessing.

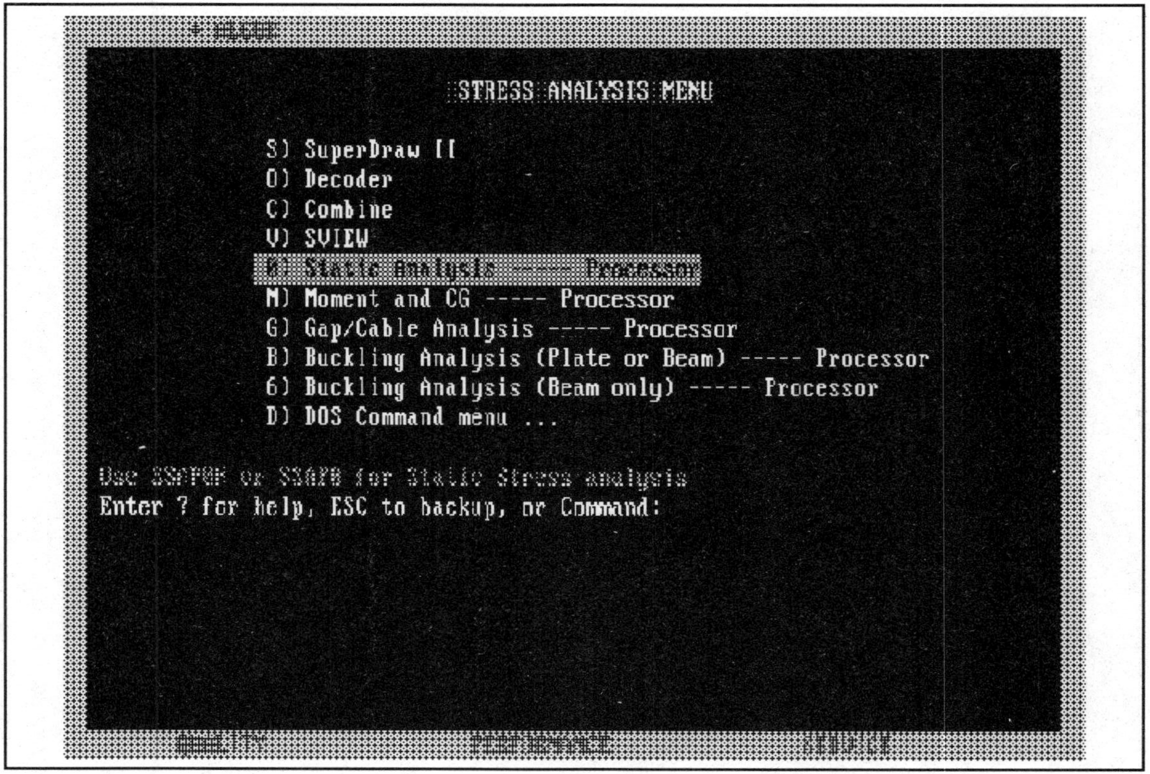

```
               :::STRESS::ANALYSIS::MENU::

      S) SuperDraw II
      D) Decoder
      C) Combine
      V) SVIEW
      U) Static Analysis      Processors
      M) Moment and CG ----- Processor
      G) Gap/Cable Analysis ----- Processor
      B) Buckling Analysis (Plate or Beam) ---- Processor
      6) Buckling Analysis (Beam only) ---- Processor
      D) DOS Command menu ...


Use SPACE or SHIFT for Static Stress analysis
Enter ? for help, ESC to backup, or Command:
```

Figure 6.7   STRESS ANALYSIS menu.

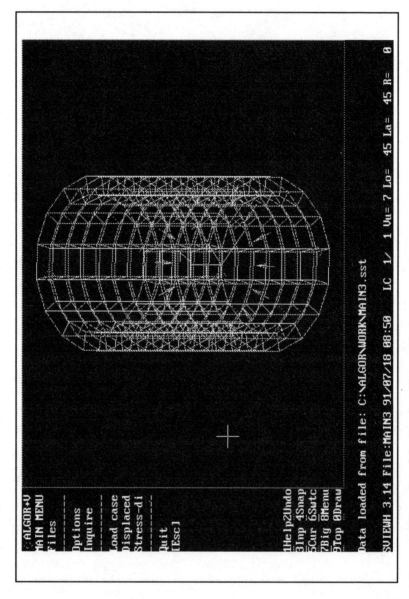

Figure 6.8 View of model after decoding and before processing.

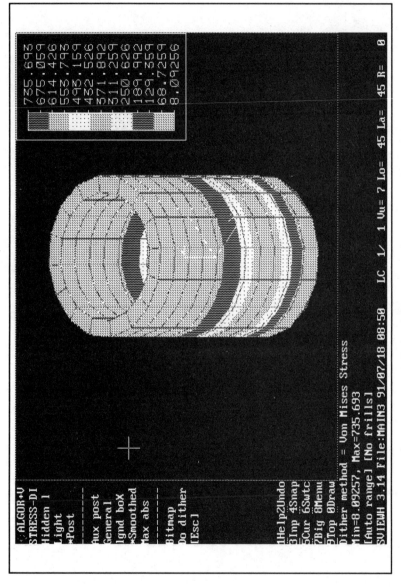

Figure 6.9 External view of cylinder stress levels.

141

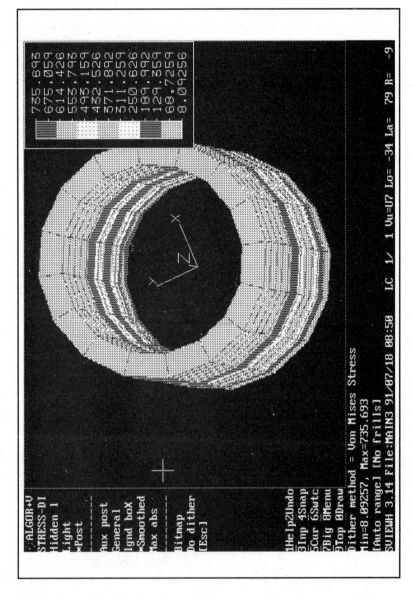

Figure 6.10 View of cylinder internal stress levels.

The gear part of the analysis was handled in much the same way as the cylindrical mating part. Initially, three concentric circles were drawn and meshed as shown in Figs. 6.11 and 6.12. The diameters of the circles are 6.309997, 6.489996, and 6.669996 inches.

Figure 6.13 illustrates the final model before processing. Note the arrows designating the surface pressure of 284 psi. Following the same path as above, the analysis shows that a radial displacement of approximately 0.00056 inch occurred. Stress and displacement values are given in Figs. 6.14 and 6.15, respectively.

The analysis confirmed that Moore could use this gear on the cylinder and that it would handle the 90 ft·lb of torque to which it would be subjected. That torque will be transmitted by the outside ring to the inner ring and will not slip. The ALGOR program also determined that the stresses, according to the von Mises criterion, were 5400 psi—very low for the type 4140 steel that would be used. It turned out that the interference fit was 0.0015 inch. This design and the numbers obtained from the analysis mean that Moore could relax manufacturing tolerances from those that had originally been anticipated, so they wouldn't be ridiculously high.

6.4 Heat Transfer in a Mold

Problems often arise in determining the optimum parameters (such as proper amount and temperature of the cooling water) to use for a particular mold application. Certainly, the use of mock-ups will cost time, personnel, and therefore money. The following problem analyzes a two-dimensional slice of a portion of a mold in order to look at temperature distributions as a function of spacing of flow passages and the number of flow passages. After the thermal analysis is completed, the temperatures at each nodal point are imputed to the identical structural model to determine the steady state stress distribution and deflection of the mold

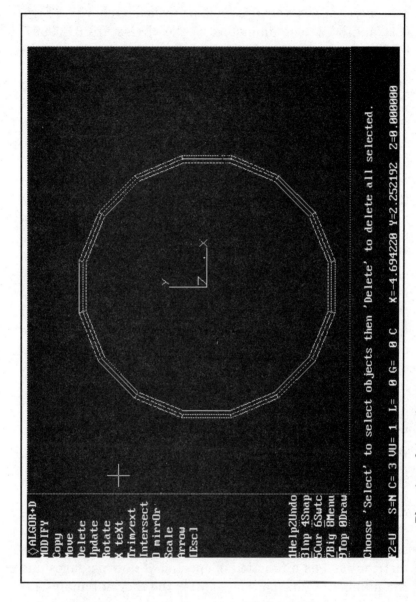

Figure 6.11 Plan view of outer ring gear.

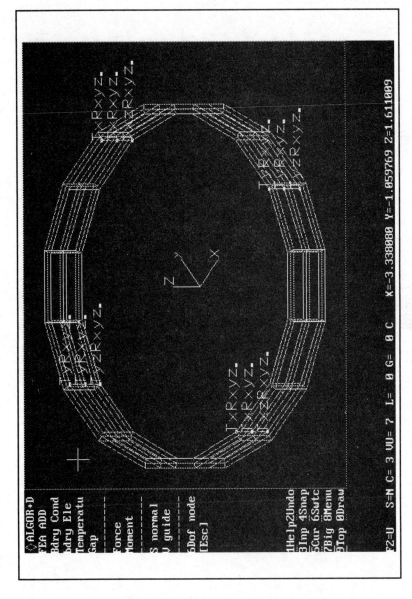

Figure 6.12 Three-dimensional view of ring gear with restraints.

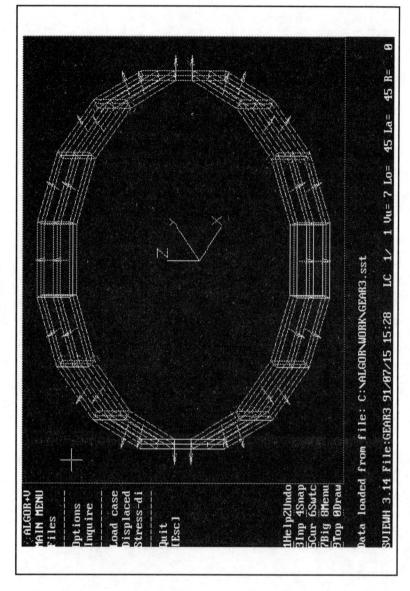

Figure 6.13 View of ring gear after preprocessing.

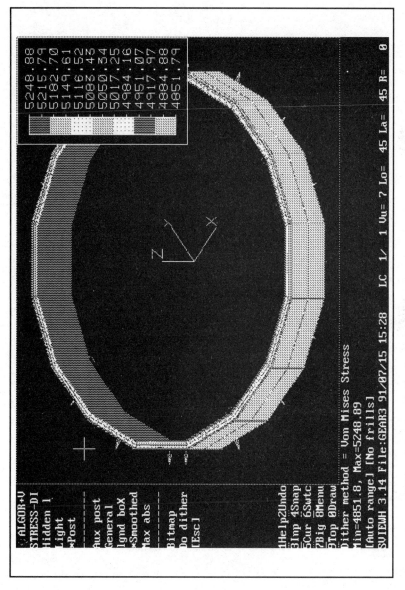

Figure 6.14 von Mises stress levels in ring gear.

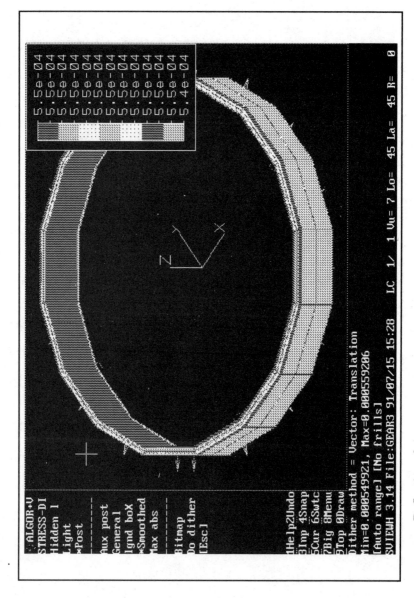

Figure 6.15 Deflection of ring gear.

cross section. The analysis could be done in full three-dimensions and for a time-based temperature loading. For simplicity, steady state loading is assumed.

Figure 6.16 illustrates the model of the half cross section of a mold. Overall dimensions are 5 inches width and 5 inches in height. Wall thickness is 0.25 inch. Material is 0.5 percent Carbon Steel.

This model has five water passages. The internal wall temperature is 450°F. It is assumed that the internal water temperature is constant at 70°F and that there exists convection from the outside surfaces into an ambient of 85°F. There are 2208 two-dimensional plate elements comprising this model. Model generation was accomplished within the ALGOR SuperDraw II package and meshed automatically by the Supergen module. This program is especially useful if there are many circles, arcs, or other items that would make regular meshing techniques time-consuming.

After preprocessing the model through the decoder, the problem is solved using the steady state heat transfer module. Output results are given for temperature levels in Fig. 6.17.

The structural model shown in Fig. 6.18 is identical to the thermal model, differing only in restraints and element type. For the structural model a two-dimensional plane stress element is used. Each nodal temperature in the structural model is imputed from the thermal model by a translation program within the ALGOR software called ADVANCE. To solve for the stresses, the linear stress module is used. Figure 6.18 shows the structural model with restraint points (basically one at each of the major corners), while Fig. 6.19 shows the deformed cross section due to temperature gradients across the model. These figures are followed by the stress distribution and deflection plots. Figs. 6.20 and 6.21, respectively.

Deflections and stress levels are high for this model. Is it possible to reduce the stress levels and deflections by adding more cooling passages? The next model shown is

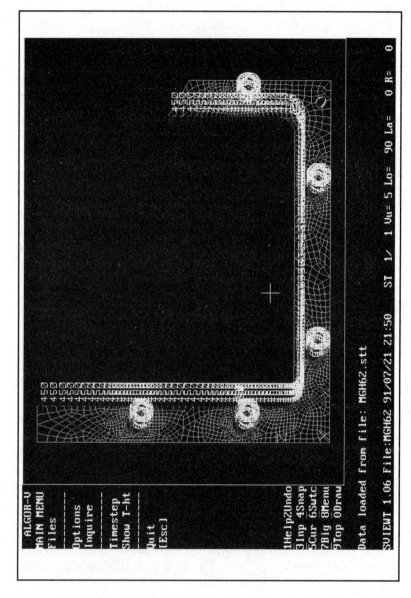

Figure 6.16 Mold cross section with five water passages.

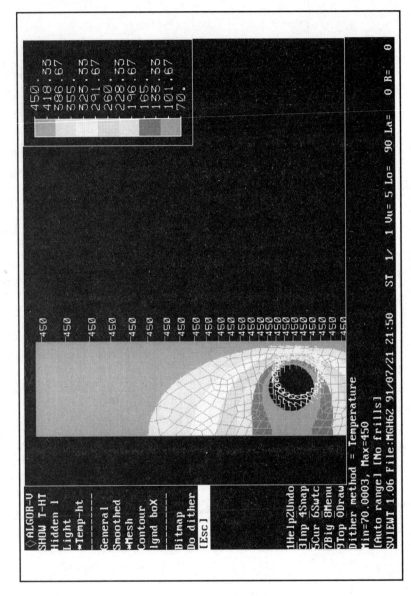

Figure 6.17 Temperature distribution—five water passage model.

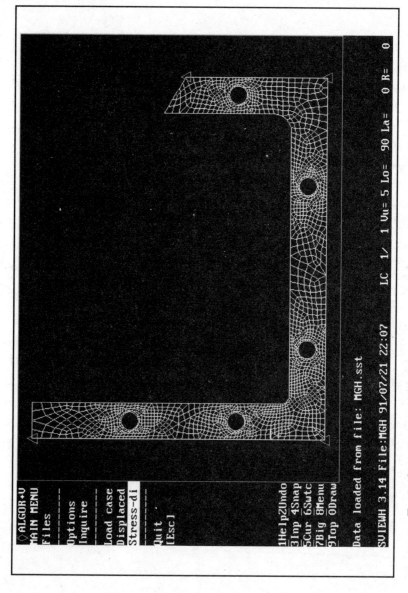

Figure 6.18 Equivalent structural model.

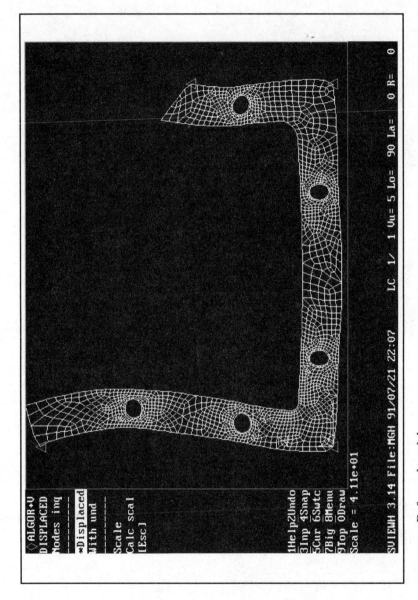

Figure 6.19 Deformed model.

153

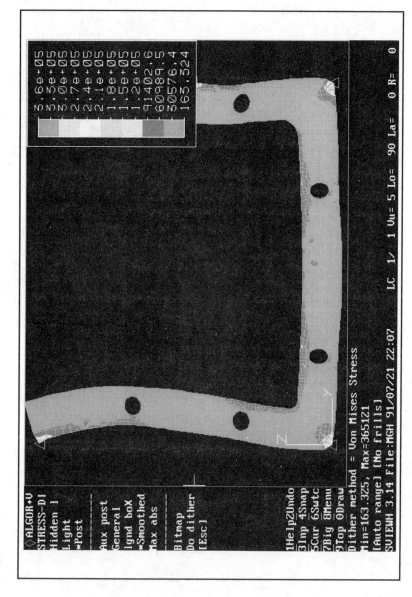

Figure 6.20 Stress distribution in mold.

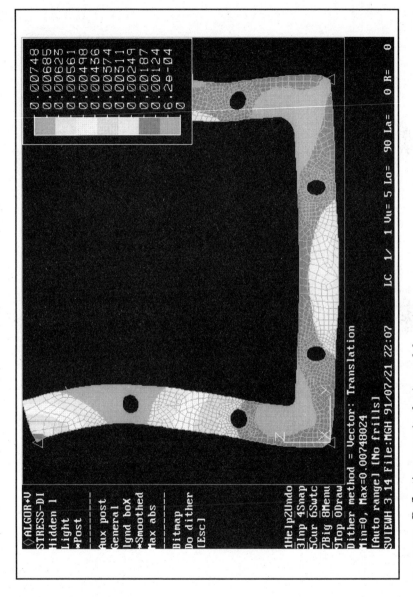

Figure 6.21 Deflection magnitudes in model.

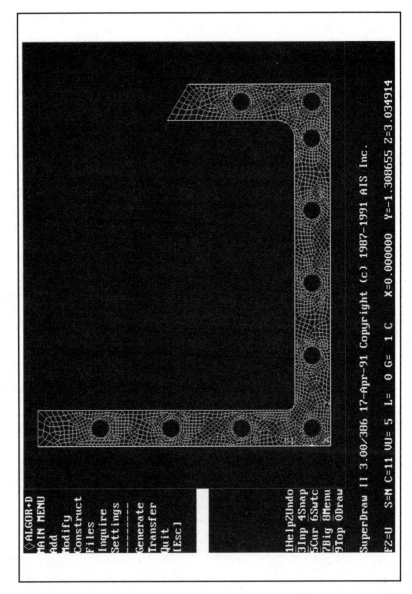

Figure 6.22 Model with additional cooling passages.

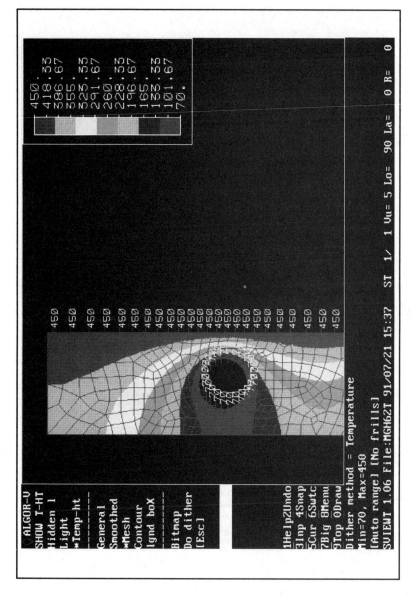

Figure 6.23 Temperature distribution in modified model.

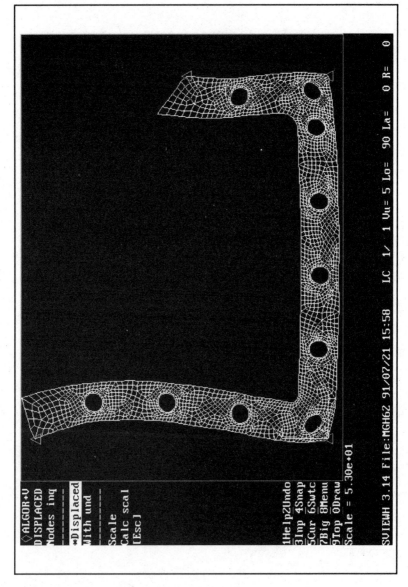

Figure 6.24 Deformed shape of new model.

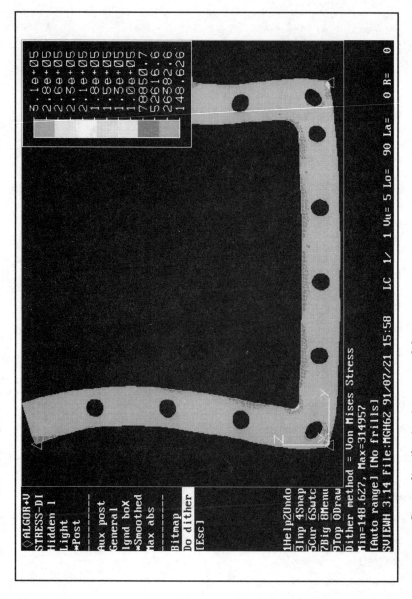

Figure 6.25 Stress distribution in new model.

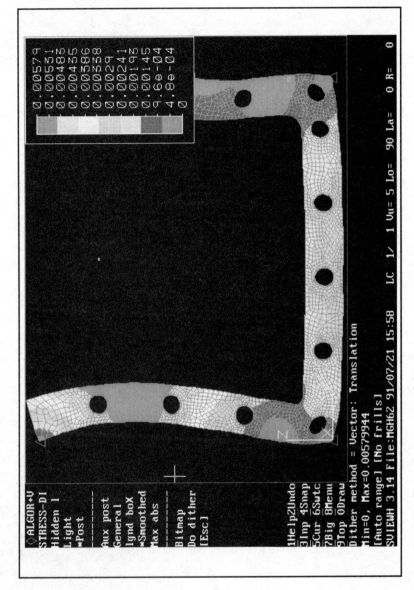

Figure 6.26 Deflections in new model.

dimensionally the same as the above problem with the addition of several cooling passages. Figure 6.22 illustrates the new configuration.

Figure 6.23 shows a close-up of the temperature distribution and is offered for comparison with Fig. 6.17. Note that the temperature gradients are not quite as large as in Fig. 6.17. Temperature boundary conditions for this new model and the old model are identical, as are material properties.

The equivalent structural model and deformed shape are shown in Fig. 6.24. The resulting stress distribution and the deflection magnitudes are given in Figs. 6.25 and 6.26, respectively.

Note that the stress levels for the model with fewer water passages are approximately 16 percent higher than for the model with ten water passages. The deflected shapes between the two models are similar with deflections for the five water passage model approximately 29 percent higher.

7

FEA in Automotive Support Industries

7.1 Introduction

This chapter examines the applications of FEA in the automotive industry. Today, the use of FEA in this arena is commonplace. In fact the use of FEA is almost a requirement, since minimizing weight while maintaining strength and overall performance of the automobile is of prime concern. In addition, the industry faces increasing global competition and strict government regulations. The goals of the automobile manufacturer are to provide safety, durability, and economy in its products. This requires the streamlining of the manufacturing and engineering processes in order to bring advanced technologies to fixed design requirements. FEA is not limited to structural applications but has great appeal in the areas of heat transfer, noise, and fluid flow. Consider, for example, the design of a muffler which requires a combined thermal-stress analysis. Hot exhaust gases produce thermal stress that can lead to early failure. All systems within an automobile can benefit from finite element analysis—everything from engine blocks, transmission housings, and suspensions to wheels, pistons, and so on. Dynamics plays a very important role in automotive design. As vehicles are designed to be lighter and smaller, suspension design is critical, since automobiles that weigh less are more sensitive to variations in shock and tire characteristics and disc rotor performance.

The first problem examined is that of the strength of a wheel hub that has applications in vehicles such as those of the four-wheel variety. The second example looks at another wheel hub application designed for minimum stress and maximum life.

The third example in this chapter looks at the two-dimensional flow around an automobile body. The velocity vectors and pressure distributions are shown and compared for two sample design cases.

7.2   Background

Warn Industries[28] is a small (400 employees), family-owned company formed by Arthur Warn who, in 1948, invented the locking hub. This first design was used to disengage the four-wheel drive on surplus military jeeps. In those days, locking was accomplished by unbolting one hub and replacing it with a different hub. The design of locking hubs has progressed quite far since the late 1940s.

Warn's engineering department has been working on a vacuum locking hub system for some years. In this system, engine vacuum is used to actuate a locking hub. The action can be controlled electrically, and the system is easily integrated with the popular "shift-on-the-fly" four-wheel drive system. One of the problems to be solved with this system is how to get the vacuum into the wheel end. In order to do this, it was found necessary to breech the axle near the point where stresses would be high: the intersection of the spindle and the hub.

The particular vehicle program for which this analysis was performed has been ongoing for just over two years. Warn has been dealing with Ssang Yong Motors in South Korea to retrofit this system to two existing four-wheel-drive vehicles. One of the installations was fairly straight-forward. The other required that a hole be drilled through the spindle. This hole raised some concerns about stress levels and was the basis for this analysis.

This section of the chapter contains analyses of three models: SPINDLE, SPINDLE0, and SPINDLE9. SPINDLE is the baseline model with no hole. SPINDLE0 has the hole aligned with the loading and SPINDLE9 has the hole 90 inches from the loads.

7.3   Description

The spindle on a front axle of a vehicle is one of the key components in transferring vehicle weight to the wheels and tires. The particular spindle analyzed in this chapter can be thought of as a circular disk with a cylinder attached at its center. There is a hole running through the middle for the axle. The disk part is rigidly bolted to the steering knuckle, while the cylinder supports the roller bearings for the wheel. Load data for the finite element models was esti-mated from the vehicle weight and length specifications.

The vacuum system required that a 4 mm hole be drilled through the base of the spindle. Since the spindle is basi-cally a cantilevered cylinder, this placed the hole in the most highly stressed region of the part.

The primary purpose of this analysis was to investigate the percentage increase in stress level caused by adding the hole. Since the loading was estimated, and could not be obtained exactly from the manufacturer, it was felt that try-ing to get an exact increase in stress would be futile. Therefore, the results presented herein represent relative variations of stress and deflection.

The baseline model contained 420 elements and the mod-els with the hole had 529 elements. The analysis was run with ALGOR's Linear Static Stress package. All drafting work was done using CADKEY and the geometry was imported using a .CDL file.

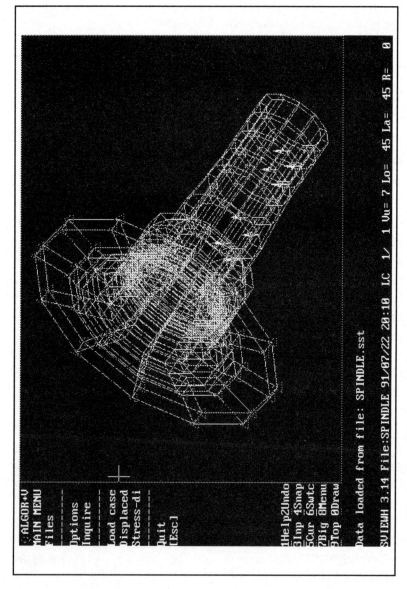

Figure 7.1 Wireframe model of spindle and hub.

The modeling process was fairly straightforward. A baseline model of the spindle was done to get the beginning stress values. Figure 7.1 illustrates the wireframe model with loading as drawn in SuperDraw II.

The following information gives the geometry and other pertinent information for the SPINDLE problem. This data can be converted and used in almost any other finite element program.

1**** Algor (c) Linear Stress Analysis SSAP0H Rel. 02/25/91, Ver. 9.20/387

DATE: JULY 22,1991
TIME: 08:04 PM
INPUT FILE.............SPINDLE

Prepared by DECODS 1.04

1**** CONTROL INFORMATION

number of node points	(NUMNP) =	672
number of element types	(NELTYP) =	1
number of load cases	(LL) =	1
number of frequencies	(NF) =	0
geometric stiffness flag	(GEOSTF) =	0
analysis type code	(NDYN) =	0
solution mode	(MODEX) =	0
equations per block	(KEQB) =	0
weight and c.g. flag	(IWTCG) =	0
bandwidth minimization flag	(MINBND) =	0
gravitational constant	(GRAV) =	3.8640E+02

bandwidth minimization specified

1**** NODAL DATA

NODE	BOUNDARY CONDITION CODES						NODAL POINT COORDINATES			
NO.	DX	DY	DZ	RX	RY	RZ	X	Y	Z	T
1	1	1	1	1	1	1	3.306E+00	3.596E+00	-2.294E+00	0.000E+00
2	1	1	1	1	1	1	3.556E+00	3.596E+00	-2.294E+00	0.000E+00
3	1	1	1	1	1	1	3.306E+00	4.825E+00	-2.294E+00	0.000E+00
4	1	1	1	1	1	1	3.556E+00	4.825E+00	-2.294E+00	0.000E+00
5	1	1	1	1	1	1	3.306E+00	2.531E+00	-1.679E+00	0.000E+00
6	1	1	1	1	1	1	3.556E+00	2.531E+00	-1.679E+00	0.000E+00
7	1	1	1	1	1	1	3.306E+00	5.890E+00	-1.679E+00	0.000E+00

8	1	1	1	1	1	1	3.556E+00	5.890E+00	-1.679E+00	0.000E+00
9	1	1	1	1	1	1	3.306E+00	3.774E+00	-1.629E+00	0.000E+00
10	1	1	1	1	1	1	3.556E+00	3.774E+00	-1.629E+00	0.000E+00
11	1	1	1	1	1	1	3.306E+00	4.647E+00	-1.629E+00	0.000E+00
12	1	1	1	1	1	1	3.556E+00	4.647E+00	-1.629E+00	0.000E+00
13	0	0	0	1	1	1	3.233E+00	3.822E+00	-1.449E+00	0.000E+00
14	0	0	0	1	1	1	3.556E+00	3.822E+00	-1.449E+00	0.000E+00
15	0	0	0	1	1	1	3.233E+00	4.598E+00	-1.449E+00	0.000E+00
16	0	0	0	1	1	1	3.556E+00	4.598E+00	-1.449E+00	0.000E+00
17	0	0	0	1	1	1	3.233E+00	3.854E+00	-1.328E+00	0.000E+00
18	0	0	0	1	1	1	3.556E+00	3.854E+00	-1.328E+00	0.000E+00
19	0	0	0	1	1	1	3.712E+00	3.854E+00	-1.328E+00	0.000E+00
20	0	0	0	1	1	1	3.233E+00	4.566E+00	-1.328E+00	0.000E+00
21	0	0	0	1	1	1	3.556E+00	4.566E+00	-1.328E+00	0.000E+00
22	0	0	0	1	1	1	3.712E+00	4.566E+00	-1.328E+00	0.000E+00
23	1	1	1	1	1	1	3.306E+00	3.018E+00	-1.192E+00	0.000E+00
24	1	1	1	1	1	1	3.556E+00	3.018E+00	-1.192E+00	0.000E+00
25	1	1	1	1	1	1	3.306E+00	5.402E+00	-1.192E+00	0.000E+00
26	1	1	1	1	1	1	3.556E+00	5.402E+00	-1.192E+00	0.000E+00
27	0	0	0	1	1	1	3.712E+00	3.894E+00	-1.178E+00	0.000E+00
28	0	0	0	1	1	1	3.712E+00	4.526E+00	-1.178E+00	0.000E+00
29	0	0	0	1	1	1	3.233E+00	3.150E+00	-1.061E+00	0.000E+00
30	0	0	0	1	1	1	3.556E+00	3.150E+00	-1.061E+00	0.000E+00
31	0	0	0	1	1	1	3.233E+00	5.271E+00	-1.061E+00	0.000E+00
32	0	0	0	1	1	1	3.556E+00	5.271E+00	-1.061E+00	0.000E+00
33	0	0	0	1	1	1	3.233E+00	3.935E+00	-1.029E+00	0.000E+00
34	0	0	0	1	1	1	3.556E+00	3.935E+00	-1.029E+00	0.000E+00
35	0	0	0	1	1	1	3.712E+00	3.935E+00	-1.029E+00	0.000E+00
36	0	0	0	1	1	1	3.837E+00	3.935E+00	-1.029E+00	0.000E+00
37	0	0	0	1	1	1	3.962E+00	3.935E+00	-1.029E+00	0.000E+00
38	0	0	0	1	1	1	4.030E+00	3.935E+00	-1.029E+00	0.000E+00
39	0	0	0	1	1	1	4.098E+00	3.935E+00	-1.029E+00	0.000E+00
40	0	0	0	1	1	1	4.166E+00	3.935E+00	-1.029E+00	0.000E+00
41	0	0	0	1	1	1	4.233E+00	3.935E+00	-1.029E+00	0.000E+00
42	0	0	0	1	1	1	4.347E+00	3.935E+00	-1.029E+00	0.000E+00
43	0	0	0	1	1	1	4.460E+00	3.935E+00	-1.029E+00	0.000E+00
44	0	0	0	1	1	1	4.686E+00	3.935E+00	-1.029E+00	0.000E+00
45	0	0	0	1	1	1	4.970E+00	3.935E+00	-1.029E+00	0.000E+00
46	0	0	0	1	1	1	3.233E+00	4.486E+00	-1.029E+00	0.000E+00
47	0	0	0	1	1	1	3.556E+00	4.486E+00	-1.029E+00	0.000E+00
48	0	0	0	1	1	1	3.712E+00	4.486E+00	-1.029E+00	0.000E+00
49	0	0	0	1	1	1	3.837E+00	4.486E+00	-1.029E+00	0.000E+00
50	0	0	0	1	1	1	3.962E+00	4.486E+00	-1.029E+00	0.000E+00
51	0	0	0	1	1	1	4.030E+00	4.486E+00	-1.029E+00	0.000E+00
52	0	0	0	1	1	1	4.098E+00	4.486E+00	-1.029E+00	0.000E+00
53	0	0	0	1	1	1	4.166E+00	4.486E+00	-1.029E+00	0.000E+00
54	0	0	0	1	1	1	4.233E+00	4.486E+00	-1.029E+00	0.000E+00
55	0	0	0	1	1	1	4.347E+00	4.486E+00	-1.029E+00	0.000E+00
56	0	0	0	1	1	1	4.460E+00	4.486E+00	-1.029E+00	0.000E+00
57	0	0	0	1	1	1	4.686E+00	4.486E+00	-1.029E+00	0.000E+00

58	0	0	0	1	1	1	4.970E+00	4.486E+00	-1.029E+00	0.000E+00
59	0	0	0	1	1	1	3.233E+00	3.238E+00	-9.723E-01	0.000E+00
60	0	0	0	1	1	1	3.556E+00	3.238E+00	-9.723E-01	0.000E+00
61	0	0	0	1	1	1	3.712E+00	3.238E+00	-9.723E-01	0.000E+00
62	0	0	0	1	1	1	3.233E+00	5.182E+00	-9.723E-01	0.000E+00
63	0	0	0	1	1	1	3.556E+00	5.182E+00	-9.723E-01	0.000E+00
64	0	0	0	1	1	1	3.712E+00	5.182E+00	-9.723E-01	0.000E+00
65	0	0	0	1	1	1	3.962E+00	3.967E+00	-9.068E-01	0.000E+00
66	0	0	0	1	1	1	4.030E+00	3.967E+00	-9.068E-01	0.000E+00
67	0	0	0	1	1	1	4.098E+00	3.967E+00	-9.068E-01	0.000E+00
68	0	0	0	1	1	1	4.166E+00	3.967E+00	-9.068E-01	0.000E+00
69	0	0	0	1	1	1	4.233E+00	3.967E+00	-9.068E-01	0.000E+00
70	0	0	0	1	1	1	3.962E+00	4.453E+00	-9.068E-01	0.000E+00
71	0	0	0	1	1	1	4.030E+00	4.453E+00	-9.068E-01	0.000E+00
72	0	0	0	1	1	1	4.098E+00	4.453E+00	-9.068E-01	0.000E+00
73	0	0	0	1	1	1	4.166E+00	4.453E+00	-9.068E-01	0.000E+00
74	0	0	0	1	1	1	4.233E+00	4.453E+00	-9.068E-01	0.000E+00
75	0	0	0	1	1	1	4.970E+00	3.974E+00	-8.813E-01	0.000E+00
76	0	0	0	1	1	1	4.970E+00	4.446E+00	-8.813E-01	0.000E+00
77	0	0	0	1	1	1	3.712E+00	3.348E+00	-8.627E-01	0.000E+00
78	0	0	0	1	1	1	3.712E+00	5.073E+00	-8.627E-01	0.000E+00
79	0	0	0	1	1	1	3.233E+00	4.000E+00	-7.848E-01	0.000E+00
80	0	0	0	1	1	1	3.556E+00	4.000E+00	-7.848E-01	0.000E+00
81	0	0	0	1	1	1	3.712E+00	4.000E+00	-7.848E-01	0.000E+00
82	0	0	0	1	1	1	3.837E+00	4.000E+00	-7.848E-01	0.000E+00
83	0	0	0	1	1	1	3.962E+00	4.000E+00	-7.848E-01	0.000E+00
84	0	0	0	1	1	1	4.030E+00	4.000E+00	-7.848E-01	0.000E+00
85	0	0	0	1	1	1	4.098E+00	4.000E+00	-7.848E-01	0.000E+00
86	0	0	0	1	1	1	4.166E+00	4.000E+00	-7.848E-01	0.000E+00
87	0	0	0	1	1	1	4.233E+00	4.000E+00	-7.848E-01	0.000E+00
88	0	0	0	1	1	1	3.233E+00	4.420E+00	-7.848E-01	0.000E+00
89	0	0	0	1	1	1	3.556E+00	4.420E+00	-7.848E-01	0.000E+00
90	0	0	0	1	1	1	3.712E+00	4.420E+00	-7.848E-01	0.000E+00
91	0	0	0	1	1	1	3.837E+00	4.420E+00	-7.848E-01	0.000E+00
92	0	0	0	1	1	1	3.962E+00	4.420E+00	-7.848E-01	0.000E+00
93	0	0	0	1	1	1	4.030E+00	4.420E+00	-7.848E-01	0.000E+00
94	0	0	0	1	1	1	4.098E+00	4.420E+00	-7.848E-01	0.000E+00
95	0	0	0	1	1	1	4.166E+00	4.420E+00	-7.848E-01	0.000E+00
96	0	0	0	1	1	1	4.233E+00	4.420E+00	-7.848E-01	0.000E+00
97	0	0	0	1	1	1	5.095E+00	4.006E+00	-7.606E-01	0.000E+00
98	0	0	0	1	1	1	5.698E+00	4.006E+00	-7.606E-01	0.000E+00
99	0	0	0	1	1	1	5.095E+00	4.414E+00	-7.606E-01	0.000E+00
100	0	0	0	1	1	1	5.698E+00	4.414E+00	-7.606E-01	0.000E+00
101	0	0	0	1	1	1	3.233E+00	3.457E+00	-7.531E-01	0.000E+00
102	0	0	0	1	1	1	3.556E+00	3.457E+00	-7.531E-01	0.000E+00
103	0	0	0	1	1	1	3.712E+00	3.457E+00	-7.531E-01	0.000E+00
104	0	0	0	1	1	1	3.837E+00	3.457E+00	-7.531E-01	0.000E+00
105	0	0	0	1	1	1	3.962E+00	3.457E+00	-7.531E-01	0.000E+00
106	0	0	0	1	1	1	4.030E+00	3.457E+00	-7.531E-01	0.000E+00
107	0	0	0	1	1	1	4.098E+00	3.457E+00	-7.531E-01	0.000E+00

108	0	0	0	1	1	1	4.166E+00	3.457E+00	-7.531E-01	0.000E+00
109	0	0	0	1	1	1	4.233E+00	3.457E+00	-7.531E-01	0.000E+00
110	0	0	0	1	1	1	4.347E+00	3.457E+00	-7.531E-01	0.000E+00
111	0	0	0	1	1	1	4.460E+00	3.457E+00	-7.531E-01	0.000E+00
112	0	0	0	1	1	1	4.686E+00	3.457E+00	-7.531E-01	0.000E+00
113	0	0	0	1	1	1	4.970E+00	3.457E+00	-7.531E-01	0.000E+00
114	0	0	0	1	1	1	3.233E+00	4.963E+00	-7.531E-01	0.000E+00
115	0	0	0	1	1	1	3.556E+00	4.963E+00	-7.531E-01	0.000E+00
116	0	0	0	1	1	1	3.712E+00	4.963E+00	-7.531E-01	0.000E+00
117	0	0	0	1	1	1	3.837E+00	4.963E+00	-7.531E-01	0.000E+00
118	0	0	0	1	1	1	3.962E+00	4.963E+00	-7.531E-01	0.000E+00
119	0	0	0	1	1	1	4.030E+00	4.963E+00	-7.531E-01	0.000E+00
120	0	0	0	1	1	1	4.098E+00	4.963E+00	-7.531E-01	0.000E+00
121	0	0	0	1	1	1	4.166E+00	4.963E+00	-7.531E-01	0.000E+00
122	0	0	0	1	1	1	4.233E+00	4.963E+00	-7.531E-01	0.000E+00
123	0	0	0	1	1	1	4.347E+00	4.963E+00	-7.531E-01	0.000E+00
124	0	0	0	1	1	1	4.460E+00	4.963E+00	-7.531E-01	0.000E+00
125	0	0	0	1	1	1	4.686E+00	4.963E+00	-7.531E-01	0.000E+00
126	0	0	0	1	1	1	4.970E+00	4.963E+00	-7.531E-01	0.000E+00
127	0	0	0	1	1	1	4.347E+00	4.017E+00	-7.216E-01	0.000E+00
128	0	0	0	1	1	1	4.347E+00	4.404E+00	-7.216E-01	0.000E+00
129	0	0	0	1	1	1	6.433E+00	4.023E+00	-7.004E-01	0.000E+00
130	0	0	0	1	1	1	6.433E+00	4.398E+00	-7.004E-01	0.000E+00
131	0	0	0	1	1	1	3.962E+00	3.546E+00	-6.638E-01	0.000E+00
132	0	0	0	1	1	1	4.030E+00	3.546E+00	-6.638E-01	0.000E+00
133	0	0	0	1	1	1	4.098E+00	3.546E+00	-6.638E-01	0.000E+00
134	0	0	0	1	1	1	4.166E+00	3.546E+00	-6.638E-01	0.000E+00
135	0	0	0	1	1	1	4.233E+00	3.546E+00	-6.638E-01	0.000E+00
136	0	0	0	1	1	1	6.859E+00	4.032E+00	-6.655E-01	0.000E+00
137	0	0	0	1	1	1	7.285E+00	4.032E+00	-6.655E-01	0.000E+00
138	0	0	0	1	1	1	7.711E+00	4.032E+00	-6.655E-01	0.000E+00
139	0	0	0	1	1	1	8.139E+00	4.032E+00	-6.655E-01	0.000E+00
140	0	0	0	1	1	1	6.859E+00	4.389E+00	-6.655E-01	0.000E+00
141	0	0	0	1	1	1	7.285E+00	4.389E+00	-6.655E-01	0.000E+00
142	0	0	0	1	1	1	7.711E+00	4.389E+00	-6.655E-01	0.000E+00
143	0	0	0	1	1	1	8.139E+00	4.389E+00	-6.655E-01	0.000E+00
144	0	0	0	1	1	1	3.962E+00	4.874E+00	-6.638E-01	0.000E+00
145	0	0	0	1	1	1	4.030E+00	4.874E+00	-6.638E-01	0.000E+00
146	0	0	0	1	1	1	4.098E+00	4.874E+00	-6.638E-01	0.000E+00
147	0	0	0	1	1	1	4.166E+00	4.874E+00	-6.638E-01	0.000E+00
148	0	0	0	1	1	1	4.233E+00	4.874E+00	-6.638E-01	0.000E+00
149	0	0	0	1	1	1	4.460E+00	4.034E+00	-6.584E-01	0.000E+00
150	0	0	0	1	1	1	4.460E+00	4.387E+00	-6.584E-01	0.000E+00
151	0	0	0	1	1	1	4.970E+00	3.565E+00	-6.452E-01	0.000E+00
152	0	0	0	1	1	1	4.970E+00	4.855E+00	-6.452E-01	0.000E+00
153	1	1	1	1	1	1	3.306E+00	1.916E+00	-6.147E-01	0.000E+00
154	1	1	1	1	1	1	3.556E+00	1.916E+00	-6.147E-01	0.000E+00
155	1	1	1	1	1	1	3.306E+00	6.504E+00	-6.147E-01	0.000E+00
156	1	1	1	1	1	1	3.556E+00	6.504E+00	-6.147E-01	0.000E+00
157	0	0	0	1	1	1	3.233E+00	3.636E+00	-5.745E-01	0.000E+00

158	0	0	0	1	1	1	3.556E+00	3.636E+00	-5.745E-01	0.000E+00
159	0	0	0	1	1	1	3.712E+00	3.636E+00	-5.745E-01	0.000E+00
160	0	0	0	1	1	1	3.837E+00	3.636E+00	-5.745E-01	0.000E+00
161	0	0	0	1	1	1	3.962E+00	3.636E+00	-5.745E-01	0.000E+00
162	0	0	0	1	1	1	4.030E+00	3.636E+00	-5.745E-01	0.000E+00
163	0	0	0	1	1	1	4.098E+00	3.636E+00	-5.745E-01	0.000E+00
164	0	0	0	1	1	1	4.166E+00	3.636E+00	-5.745E-01	0.000E+00
165	0	0	0	1	1	1	4.233E+00	3.636E+00	-5.745E-01	0.000E+00
166	0	0	0	1	1	1	3.233E+00	4.785E+00	-5.745E-01	0.000E+00
167	0	0	0	1	1	1	3.556E+00	4.785E+00	-5.745E-01	0.000E+00
168	0	0	0	1	1	1	3.712E+00	4.785E+00	-5.745E-01	0.000E+00
169	0	0	0	1	1	1	3.837E+00	4.785E+00	-5.745E-01	0.000E+00
170	0	0	0	1	1	1	3.962E+00	4.785E+00	-5.745E-01	0.000E+00
171	0	0	0	1	1	1	4.030E+00	4.785E+00	-5.745E-01	0.000E+00
172	0	0	0	1	1	1	4.098E+00	4.785E+00	-5.745E-01	0.000E+00
173	0	0	0	1	1	1	4.166E+00	4.785E+00	-5.745E-01	0.000E+00
174	0	0	0	1	1	1	4.233E+00	4.785E+00	-5.745E-01	0.000E+00
175	0	0	0	1	1	1	5.095E+00	3.653E+00	-5.568E-01	0.000E+00
176	0	0	0	1	1	1	5.698E+00	3.653E+00	-5.568E-01	0.000E+00
177	0	0	0	1	1	1	5.095E+00	4.767E+00	-5.568E-01	0.000E+00
178	0	0	0	1	1	1	5.698E+00	4.767E+00	-5.568E-01	0.000E+00
179	0	0	0	1	1	1	4.686E+00	4.068E+00	-5.324E-01	0.000E+00
180	0	0	0	1	1	1	4.970E+00	4.068E+00	-5.324E-01	0.000E+00
181	0	0	0	1	1	1	5.095E+00	4.068E+00	-5.324E-01	0.000E+00
182	0	0	0	1	1	1	5.698E+00	4.068E+00	-5.324E-01	0.000E+00
183	0	0	0	1	1	1	6.433E+00	4.068E+00	-5.324E-01	0.000E+00
184	0	0	0	1	1	1	6.859E+00	4.068E+00	-5.324E-01	0.000E+00
185	0	0	0	1	1	1	7.285E+00	4.068E+00	-5.324E-01	0.000E+00
186	0	0	0	1	1	1	7.711E+00	4.068E+00	-5.324E-01	0.000E+00
187	0	0	0	1	1	1	8.139E+00	4.068E+00	-5.324E-01	0.000E+00
188	0	0	0	1	1	1	4.686E+00	4.353E+00	-5.324E-01	0.000E+00
189	0	0	0	1	1	1	4.970E+00	4.353E+00	-5.324E-01	0.000E+00
190	0	0	0	1	1	1	5.095E+00	4.353E+00	-5.324E-01	0.000E+00
191	0	0	0	1	1	1	5.698E+00	4.353E+00	-5.324E-01	0.000E+00
192	0	0	0	1	1	1	6.433E+00	4.353E+00	-5.324E-01	0.000E+00
193	0	0	0	1	1	1	6.859E+00	4.353E+00	-5.324E-01	0.000E+00
194	0	0	0	1	1	1	7.285E+00	4.353E+00	-5.324E-01	0.000E+00
195	0	0	0	1	1	1	7.711E+00	4.353E+00	-5.324E-01	0.000E+00
196	0	0	0	1	1	1	8.139E+00	4.353E+00	-5.324E-01	0.000E+00
197	0	0	0	1	1	1	4.347E+00	3.682E+00	-5.283E-01	0.000E+00
198	0	0	0	1	1	1	4.347E+00	4.738E+00	-5.283E-01	0.000E+00
199	0	0	0	1	1	1	6.433E+00	3.697E+00	-5.127E-01	0.000E+00
200	0	0	0	1	1	1	6.433E+00	4.723E+00	-5.127E-01	0.000E+00
201	0	0	0	1	1	1	6.859E+00	3.723E+00	-4.872E-01	0.000E+00
202	0	0	0	1	1	1	7.285E+00	3.723E+00	-4.872E-01	0.000E+00
203	0	0	0	1	1	1	7.711E+00	3.723E+00	-4.872E-01	0.000E+00
204	0	0	0	1	1	1	8.139E+00	3.723E+00	-4.872E-01	0.000E+00
205	0	0	0	1	1	1	6.859E+00	4.697E+00	-4.872E-01	0.000E+00
206	0	0	0	1	1	1	7.285E+00	4.697E+00	-4.872E-01	0.000E+00
207	0	0	0	1	1	1	7.711E+00	4.697E+00	-4.872E-01	0.000E+00

208	0	0	0	1	1	1	8.139E+00	4.697E+00	-4.872E-01	0.000E+00
209	0	0	0	1	1	1	4.460E+00	3.728E+00	-4.820E-01	0.000E+00
210	0	0	0	1	1	1	4.460E+00	4.692E+00	-4.820E-01	0.000E+00
211	1	1	1	1	1	1	3.306E+00	2.582E+00	-4.364E-01	0.000E+00
212	1	1	1	1	1	1	3.556E+00	2.582E+00	-4.364E-01	0.000E+00
213	1	1	1	1	1	1	3.306E+00	5.839E+00	-4.364E-01	0.000E+00
214	1	1	1	1	1	1	3.556E+00	5.839E+00	-4.364E-01	0.000E+00
215	0	0	0	1	1	1	3.233E+00	2.761E+00	-3.882E-01	0.000E+00
216	0	0	0	1	1	1	3.556E+00	2.761E+00	-3.882E-01	0.000E+00
217	0	0	0	1	1	1	4.686E+00	3.820E+00	-3.897E-01	0.000E+00
218	0	0	0	1	1	1	4.970E+00	3.820E+00	-3.897E-01	0.000E+00
219	0	0	0	1	1	1	5.095E+00	3.820E+00	-3.897E-01	0.000E+00
220	0	0	0	1	1	1	5.698E+00	3.820E+00	-3.897E-01	0.000E+00
221	0	0	0	1	1	1	6.433E+00	3.820E+00	-3.897E-01	0.000E+00
222	0	0	0	1	1	1	6.859E+00	3.820E+00	-3.897E-01	0.000E+00
223	0	0	0	1	1	1	7.285E+00	3.820E+00	-3.897E-01	0.000E+00
224	0	0	0	1	1	1	7.711E+00	3.820E+00	-3.897E-01	0.000E+00
225	0	0	0	1	1	1	8.139E+00	3.820E+00	-3.897E-01	0.000E+00
226	0	0	0	1	1	1	4.686E+00	4.600E+00	-3.897E-01	0.000E+00
227	0	0	0	1	1	1	4.970E+00	4.600E+00	-3.897E-01	0.000E+00
228	0	0	0	1	1	1	5.095E+00	4.600E+00	-3.897E-01	0.000E+00
229	0	0	0	1	1	1	5.698E+00	4.600E+00	-3.897E-01	0.000E+00
230	0	0	0	1	1	1	6.433E+00	4.600E+00	-3.897E-01	0.000E+00
231	0	0	0	1	1	1	6.859E+00	4.600E+00	-3.897E-01	0.000E+00
232	0	0	0	1	1	1	7.285E+00	4.600E+00	-3.897E-01	0.000E+00
233	0	0	0	1	1	1	7.711E+00	4.600E+00	-3.897E-01	0.000E+00
234	0	0	0	1	1	1	8.139E+00	4.600E+00	-3.897E-01	0.000E+00
235	0	0	0	1	1	1	3.233E+00	5.659E+00	-3.882E-01	0.000E+00
236	0	0	0	1	1	1	3.556E+00	5.659E+00	-3.882E-01	0.000E+00
237	0	0	0	1	1	1	3.233E+00	2.882E+00	-3.559E-01	0.000E+00
238	0	0	0	1	1	1	3.556E+00	2.882E+00	-3.559E-01	0.000E+00
239	0	0	0	1	1	1	3.712E+00	2.882E+00	-3.559E-01	0.000E+00
240	0	0	0	1	1	1	3.233E+00	5.538E+00	-3.559E-01	0.000E+00
241	0	0	0	1	1	1	3.556E+00	5.538E+00	-3.559E-01	0.000E+00
242	0	0	0	1	1	1	3.712E+00	5.538E+00	-3.559E-01	0.000E+00
243	0	0	0	1	1	1	3.712E+00	3.032E+00	-3.158E-01	0.000E+00
244	0	0	0	1	1	1	3.712E+00	5.389E+00	-3.158E-01	0.000E+00
245	0	0	0	1	1	1	3.233E+00	3.181E+00	-2.756E-01	0.000E+00
246	0	0	0	1	1	1	3.556E+00	3.181E+00	-2.756E-01	0.000E+00
247	0	0	0	1	1	1	3.712E+00	3.181E+00	-2.756E-01	0.000E+00
248	0	0	0	1	1	1	3.837E+00	3.181E+00	-2.756E-01	0.000E+00
249	0	0	0	1	1	1	3.962E+00	3.181E+00	-2.756E-01	0.000E+00
250	0	0	0	1	1	1	4.030E+00	3.181E+00	-2.756E-01	0.000E+00
251	0	0	0	1	1	1	4.098E+00	3.181E+00	-2.756E-01	0.000E+00
252	0	0	0	1	1	1	4.166E+00	3.181E+00	-2.756E-01	0.000E+00
253	0	0	0	1	1	1	4.233E+00	3.181E+00	-2.756E-01	0.000E+00
254	0	0	0	1	1	1	4.347E+00	3.181E+00	-2.756E-01	0.000E+00
255	0	0	0	1	1	1	4.460E+00	3.181E+00	-2.756E-01	0.000E+00
256	0	0	0	1	1	1	4.686E+00	3.181E+00	-2.756E-01	0.000E+00
257	0	0	0	1	1	1	4.970E+00	3.181E+00	-2.756E-01	0.000E+00

258	0	0	0	1	1	1	3.233E+00	5.239E+00	-2.756E-01	0.000E+00
259	0	0	0	1	1	1	3.556E+00	5.239E+00	-2.756E-01	0.000E+00
260	0	0	0	1	1	1	3.712E+00	5.239E+00	-2.756E-01	0.000E+00
261	0	0	0	1	1	1	3.837E+00	5.239E+00	-2.756E-01	0.000E+00
262	0	0	0	1	1	1	3.962E+00	5.239E+00	-2.756E-01	0.000E+00
263	0	0	0	1	1	1	4.030E+00	5.239E+00	-2.756E-01	0.000E+00
264	0	0	0	1	1	1	4.098E+00	5.239E+00	-2.756E-01	0.000E+00
265	0	0	0	1	1	1	4.166E+00	5.239E+00	-2.756E-01	0.000E+00
266	0	0	0	1	1	1	4.233E+00	5.239E+00	-2.756E-01	0.000E+00
267	0	0	0	1	1	1	4.347E+00	5.239E+00	-2.756E-01	0.000E+00
268	0	0	0	1	1	1	4.460E+00	5.239E+00	-2.756E-01	0.000E+00
269	0	0	0	1	1	1	4.686E+00	5.239E+00	-2.756E-01	0.000E+00
270	0	0	0	1	1	1	4.970E+00	5.239E+00	-2.756E-01	0.000E+00
271	0	0	0	1	1	1	3.962E+00	3.303E+00	-2.430E-01	0.000E+00
272	0	0	0	1	1	1	4.030E+00	3.303E+00	-2.430E-01	0.000E+00
273	0	0	0	1	1	1	4.098E+00	3.303E+00	-2.430E-01	0.000E+00
274	0	0	0	1	1	1	4.166E+00	3.303E+00	-2.430E-01	0.000E+00
275	0	0	0	1	1	1	4.233E+00	3.303E+00	-2.430E-01	0.000E+00
276	0	0	0	1	1	1	3.962E+00	5.117E+00	-2.430E-01	0.000E+00
277	0	0	0	1	1	1	4.030E+00	5.117E+00	-2.430E-01	0.000E+00
278	0	0	0	1	1	1	4.098E+00	5.117E+00	-2.430E-01	0.000E+00
279	0	0	0	1	1	1	4.166E+00	5.117E+00	-2.430E-01	0.000E+00
280	0	0	0	1	1	1	4.233E+00	5.117E+00	-2.430E-01	0.000E+00
281	0	0	0	1	1	1	4.970E+00	3.329E+00	-2.361E-01	0.000E+00
282	0	0	0	1	1	1	4.970E+00	5.092E+00	-2.361E-01	0.000E+00
283	0	0	0	1	1	1	3.233E+00	3.425E+00	-2.103E-01	0.000E+00
284	0	0	0	1	1	1	3.556E+00	3.425E+00	-2.103E-01	0.000E+00
285	0	0	0	1	1	1	3.712E+00	3.425E+00	-2.103E-01	0.000E+00
286	0	0	0	1	1	1	3.837E+00	3.425E+00	-2.103E-01	0.000E+00
287	0	0	0	1	1	1	3.962E+00	3.425E+00	-2.103E-01	0.000E+00
288	0	0	0	1	1	1	4.030E+00	3.425E+00	-2.103E-01	0.000E+00
289	0	0	0	1	1	1	4.098E+00	3.425E+00	-2.103E-01	0.000E+00
290	0	0	0	1	1	1	4.166E+00	3.425E+00	-2.103E-01	0.000E+00
291	0	0	0	1	1	1	4.233E+00	3.425E+00	-2.103E-01	0.000E+00
292	0	0	0	1	1	1	3.233E+00	4.995E+00	-2.103E-01	0.000E+00
293	0	0	0	1	1	1	3.556E+00	4.995E+00	-2.103E-01	0.000E+00
294	0	0	0	1	1	1	3.712E+00	4.995E+00	-2.103E-01	0.000E+00
295	0	0	0	1	1	1	3.837E+00	4.995E+00	-2.103E-01	0.000E+00
296	0	0	0	1	1	1	3.962E+00	4.995E+00	-2.103E-01	0.000E+00
297	0	0	0	1	1	1	4.030E+00	4.995E+00	-2.103E-01	0.000E+00
298	0	0	0	1	1	1	4.098E+00	4.995E+00	-2.103E-01	0.000E+00
299	0	0	0	1	1	1	4.166E+00	4.995E+00	-2.103E-01	0.000E+00
300	0	0	0	1	1	1	4.233E+00	4.995E+00	-2.103E-01	0.000E+00
301	0	0	0	1	1	1	5.095E+00	3.450E+00	-2.038E-01	0.000E+00
302	0	0	0	1	1	1	5.698E+00	3.450E+00	-2.038E-01	0.000E+00
303	0	0	0	1	1	1	5.095E+00	4.971E+00	-2.038E-01	0.000E+00
304	0	0	0	1	1	1	5.698E+00	4.971E+00	-2.038E-01	0.000E+00
305	0	0	0	1	1	1	4.347E+00	3.489E+00	-1.934E-01	0.000E+00
306	0	0	0	1	1	1	4.347E+00	4.932E+00	-1.934E-01	0.000E+00
307	0	0	0	1	1	1	6.433E+00	3.510E+00	-1.877E-01	0.000E+00

308	0	0	0	1	1	1	6.433E+00	4.911E+00	-1.877E-01	0.000E+00
309	0	0	0	1	1	1	6.859E+00	3.545E+00	-1.783E-01	0.000E+00
310	0	0	0	1	1	1	7.285E+00	3.545E+00	-1.783E-01	0.000E+00
311	0	0	0	1	1	1	7.711E+00	3.545E+00	-1.783E-01	0.000E+00
312	0	0	0	1	1	1	8.139E+00	3.545E+00	-1.783E-01	0.000E+00
313	0	0	0	1	1	1	4.460E+00	3.552E+00	-1.764E-01	0.000E+00
314	0	0	0	1	1	1	4.460E+00	4.869E+00	-1.764E-01	0.000E+00
315	0	0	0	1	1	1	6.859E+00	4.876E+00	-1.783E-01	0.000E+00
316	0	0	0	1	1	1	7.285E+00	4.876E+00	-1.783E-01	0.000E+00
317	0	0	0	1	1	1	7.711E+00	4.876E+00	-1.783E-01	0.000E+00
318	0	0	0	1	1	1	8.139E+00	4.876E+00	-1.783E-01	0.000E+00
319	0	0	0	1	1	1	4.686E+00	3.678E+00	-1.427E-01	0.000E+00
320	0	0	0	1	1	1	4.970E+00	3.678E+00	-1.427E-01	0.000E+00
321	0	0	0	1	1	1	5.095E+00	3.678E+00	-1.427E-01	0.000E+00
322	0	0	0	1	1	1	5.698E+00	3.678E+00	-1.427E-01	0.000E+00
323	0	0	0	1	1	1	6.433E+00	3.678E+00	-1.427E-01	0.000E+00
324	0	0	0	1	1	1	6.859E+00	3.678E+00	-1.427E-01	0.000E+00
325	0	0	0	1	1	1	7.285E+00	3.678E+00	-1.427E-01	0.000E+00
326	0	0	0	1	1	1	7.711E+00	3.678E+00	-1.427E-01	0.000E+00
327	0	0	0	1	1	1	8.139E+00	3.678E+00	-1.427E-01	0.000E+00
328	0	0	0	1	1	1	4.686E+00	4.743E+00	-1.427E-01	0.000E+00
329	0	0	0	1	1	1	4.970E+00	4.743E+00	-1.427E-01	0.000E+00
330	0	0	0	1	1	1	5.095E+00	4.743E+00	-1.427E-01	0.000E+00
331	0	0	0	1	1	1	5.698E+00	4.743E+00	-1.427E-01	0.000E+00
332	0	0	0	1	1	1	6.433E+00	4.743E+00	-1.427E-01	0.000E+00
333	0	0	0	1	1	1	6.859E+00	4.743E+00	-1.427E-01	0.000E+00
334	0	0	0	1	1	1	7.285E+00	4.743E+00	-1.427E-01	0.000E+00
335	0	0	0	1	1	1	7.711E+00	4.743E+00	-1.427E-01	0.000E+00
336	0	0	0	1	1	1	8.139E+00	4.743E+00	-1.427E-01	0.000E+00
337	0	0	0	1	1	1	4.686E+00	3.678E+00	1.427E-01	0.000E+00
338	0	0	0	1	1	1	4.970E+00	3.678E+00	1.427E-01	0.000E+00
339	0	0	0	1	1	1	5.095E+00	3.678E+00	1.427E-01	0.000E+00
340	0	0	0	1	1	1	5.698E+00	3.678E+00	1.427E-01	0.000E+00
341	0	0	0	1	1	1	6.433E+00	3.678E+00	1.427E-01	0.000E+00
342	0	0	0	1	1	1	6.859E+00	3.678E+00	1.427E-01	0.000E+00
343	0	0	0	1	1	1	7.285E+00	3.678E+00	1.427E-01	0.000E+00
344	0	0	0	1	1	1	7.711E+00	3.678E+00	1.427E-01	0.000E+00
345	0	0	0	1	1	1	8.139E+00	3.678E+00	1.427E-01	0.000E̤+00
346	0	0	0	1	1	1	4.686E+00	4.743E+00	1.427E-01	0.000E+00
347	0	0	0	1	1	1	4.970E+00	4.743E+00	1.427E-01	0.000E+00
348	0	0	0	1	1	1	5.095E+00	4.743E+00	1.427E-01	0.000E+00
349	0	0	0	1	1	1	5.698E+00	4.743E+00	1.427E-01	0.000E+00
350	0	0	0	1	1	1	6.433E+00	4.743E+00	1.427E-01	0.000E+00
351	0	0	0	1	1	1	6.859E+00	4.743E+00	1.427E-01	0.000E+00
352	0	0	0	1	1	1	7.285E+00	4.743E+00	1.427E-01	0.000E+00
353	0	0	0	1	1	1	7.711E+00	4.743E+00	1.427E-01	0.000E+00
354	0	0	0	1	1	1	8.139E+00	4.743E+00	1.427E-01	0.000E+00
355	0	0	0	1	1	1	6.859E+00	3.545E+00	1.783E-01	0.000E+00
356	0	0	0	1	1	1	7.285E+00	3.545E+00	1.783E-01	0.000E+00
357	0	0	0	1	1	1	7.711E+00	3.545E+00	1.783E-01	0.000E+00
358	0	0	0	1	1	1	8.139E+00	3.545E+00	1.783E-01	0.000E+00

359	0	0	0	1	1	1	4.460E+00	3.552E+00	1.764E-01	0.000E+00
360	0	0	0	1	1	1	4.460E+00	4.869E+00	1.764E-01	0.000E+00
361	0	0	0	1	1	1	6.859E+00	4.876E+00	1.783E-01	0.000E+00
362	0	0	0	1	1	1	7.285E+00	4.876E+00	1.783E-01	0.000E+00
363	0	0	0	1	1	1	7.711E+00	4.876E+00	1.783E-01	0.000E+00
364	0	0	0	1	1	1	8.139E+00	4.876E+00	1.783E-01	0.000E+00
365	0	0	0	1	1	1	6.433E+00	3.510E+00	1.877E-01	0.000E+00
366	0	0	0	1	1	1	6.433E+00	4.911E+00	1.877E-01	0.000E+00
367	0	0	0	1	1	1	4.347E+00	3.489E+00	1.934E-01	0.000E+00
368	0	0	0	1	1	1	4.347E+00	4.932E+00	1.934E-01	0.000E+00
369	0	0	0	1	1	1	5.095E+00	3.450E+00	2.038E-01	0.000E+00
370	0	0	0	1	1	1	5.698E+00	3.450E+00	2.038E-01	0.000E+00
371	0	0	0	1	1	1	5.095E+00	4.971E+00	2.038E-01	0.000E+00
372	0	0	0	1	1	1	5.698E+00	4.971E+00	2.038E-01	0.000E+00
373	0	0	0	1	1	1	3.233E+00	3.425E+00	2.103E-01	0.000E+00
374	0	0	0	1	1	1	3.556E+00	3.425E+00	2.103E-01	0.000E+00
375	0	0	0	1	1	1	3.712E+00	3.425E+00	2.103E-01	0.000E+00
376	0	0	0	1	1	1	3.837E+00	3.425E+00	2.103E-01	0.000E+00
377	0	0	0	1	1	1	3.962E+00	3.425E+00	2.103E-01	0.000E+00
378	0	0	0	1	1	1	4.030E+00	3.425E+00	2.103E-01	0.000E+00
379	0	0	0	1	1	1	4.098E+00	3.425E+00	2.103E-01	0.000E+00
380	0	0	0	1	1	1	4.166E+00	3.425E+00	2.103E-01	0.000E+00
381	0	0	0	1	1	1	4.233E+00	3.425E+00	2.103E-01	0.000E+00
382	0	0	0	1	1	1	3.233E+00	4.995E+00	2.103E-01	0.000E+00
383	0	0	0	1	1	1	3.556E+00	4.995E+00	2.103E-01	0.000E+00
384	0	0	0	1	1	1	3.712E+00	4.995E+00	2.103E-01	0.000E+00
385	0	0	0	1	1	1	3.837E+00	4.995E+00	2.103E-01	0.000E+00
386	0	0	0	1	1	1	3.962E+00	4.995E+00	2.103E-01	0.000E+00
387	0	0	0	1	1	1	4.030E+00	4.995E+00	2.103E-01	0.000E+00
388	0	0	0	1	1	1	4.098E+00	4.995E+00	2.103E-01	0.000E+00
389	0	0	0	1	1	1	4.166E+00	4.995E+00	2.103E-01	0.000E+00
390	0	0	0	1	1	1	4.233E+00	4.995E+00	2.103E-01	0.000E+00
391	0	0	0	1	1	1	4.970E+00	3.329E+00	2.361E-01	0.000E+00
392	0	0	0	1	1	1	4.970E+00	5.092E+00	2.361E-01	0.000E+00
393	0	0	0	1	1	1	3.962E+00	3.303E+00	2.430E-01	0.000E+00
394	0	0	0	1	1	1	4.030E+00	3.303E+00	2.430E-01	0.000E+00
395	0	0	0	1	1	1	4.098E+00	3.303E+00	2.430E-01	0.000E+00
396	0	0	0	1	1	1	4.166E+00	3.303E+00	2.430E-01	0.000E+00
397	0	0	0	1	1	1	4.233E+00	3.303E+00	2.430E-01	0.000E+00
398	0	0	0	1	1	1	3.962E+00	5.117E+00	2.430E-01	0.000E+00
399	0	0	0	1	1	1	4.030E+00	5.117E+00	2.430E-01	0.000E+00
400	0	0	0	1	1	1	4.098E+00	5.117E+00	2.430E-01	0.000E+00
401	0	0	0	1	1	1	4.166E+00	5.117E+00	2.430E-01	0.000E+00
402	0	0	0	1	1	1	4.233E+00	5.117E+00	2.430E-01	0.000E+00
403	0	0	0	1	1	1	3.233E+00	3.181E+00	2.756E-01	0.000E+00
404	0	0	0	1	1	1	3.556E+00	3.181E+00	2.756E-01	0.000E+00
405	0	0	0	1	1	1	3.712E+00	3.181E+00	2.756E-01	0.000E+00
406	0	0	0	1	1	1	3.837E+00	3.181E+00	2.756E-01	0.000E+00
407	0	0	0	1	1	1	3.962E+00	3.181E+00	2.756E-01	0.000E+00
408	0	0	0	1	1	1	4.030E+00	3.181E+00	2.756E-01	0.000E+00

409	0	0	0	1	1	1	4.098E+00	3.181E+00	2.756E-01	0.000E+00
410	0	0	0	1	1	1	4.166E+00	3.181E+00	2.756E-01	0.000E+00
411	0	0	0	1	1	1	4.233E+00	3.181E+00	2.756E-01	0.000E+00
412	0	0	0	1	1	1	4.347E+00	3.181E+00	2.756E-01	0.000E+00
413	0	0	0	1	1	1	4.460E+00	3.181E+00	2.756E-01	0.000E+00
414	0	0	0	1	1	1	4.686E+00	3.181E+00	2.756E-01	0.000E+00
415	0	0	0	1	1	1	4.970E+00	3.181E+00	2.756E-01	0.000E+00
416	0	0	0	1	1	1	3.233E+00	5.239E+00	2.756E-01	0.000E+00
417	0	0	0	1	1	1	3.556E+00	5.239E+00	2.756E-01	0.000E+00
418	0	0	0	1	1	1	3.712E+00	5.239E+00	2.756E-01	0.000E+00
419	0	0	0	1	1	1	3.837E+00	5.239E+00	2.756E-01	0.000E+00
420	0	0	0	1	1	1	3.962E+00	5.239E+00	2.756E-01	0.000E+00
421	0	0	0	1	1	1	4.030E+00	5.239E+00	2.756E-01	0.000E+00
422	0	0	0	1	1	1	4.098E+00	5.239E+00	2.756E-01	0.000E+00
423	0	0	0	1	1	1	4.166E+00	5.239E+00	2.756E-01	0.000E+00
424	0	0	0	1	1	1	4.233E+00	5.239E+00	2.756E-01	0.000E+00
425	0	0	0	1	1	1	4.347E+00	5.239E+00	2.756E-01	0.000E+00
426	0	0	0	1	1	1	4.460E+00	5.239E+00	2.756E-01	0.000E+00
427	0	0	0	1	1	1	4.686E+00	5.239E+00	2.756E-01	0.000E+00
428	0	0	0	1	1	1	4.970E+00	5.239E+00	2.756E-01	0.000E+00
429	0	0	0	1	1	1	3.712E+00	3.032E+00	3.158E-01	0.000E+00
430	0	0	0	1	1	1	3.712E+00	5.389E+00	3.158E-01	0.000E+00
431	0	0	0	1	1	1	3.233E+00	2.882E+00	3.559E-01	0.000E+00
432	0	0	0	1	1	1	3.556E+00	2.882E+00	3.559E-01	0.000E+00
433	0	0	0	1	1	1	3.712E+00	2.882E+00	3.559E-01	0.000E+00
434	0	0	0	1	1	1	3.233E+00	5.538E+00	3.559E-01	0.000E+00
435	0	0	0	1	1	1	3.556E+00	5.538E+00	3.559E-01	0.000E+00
436	0	0	0	1	1	1	3.712E+00	5.538E+00	3.559E-01	0.000E+00
437	0	0	0	1	1	1	3.233E+00	2.761E+00	3.882E-01	0.000E+00
438	0	0	0	1	1	1	3.556E+00	2.761E+00	3.882E-01	0.000E+00
439	0	0	0	1	1	1	4.686E+00	3.820E+00	3.897E-01	0.000E+00
440	0	0	0	1	1	1	4.970E+00	3.820E+00	3.897E-01	0.000E+00
441	0	0	0	1	1	1	5.095E+00	3.820E+00	3.897E-01	0.000E+00
442	0	0	0	1	1	1	5.698E+00	3.820E+00	3.897E-01	0.000E+00
443	0	0	0	1	1	1	6.433E+00	3.820E+00	3.897E-01	0.000E+00
444	0	0	0	1	1	1	6.859E+00	3.820E+00	3.897E-01	0.000E+00
445	0	0	0	1	1	1	7.285E+00	3.820E+00	3.897E-01	0.000E+00
446	0	0	0	1	1	1	7.711E+00	3.820E+00	3.897E-01	0.000E+00
447	0	0	0	1	1	1	8.139E+00	3.820E+00	3.897E-01	0.000E+00
448	0	0	0	1	1	1	4.686E+00	4.600E+00	3.897E-01	0.000E+00
449	0	0	0	1	1	1	4.970E+00	4.600E+00	3.897E-01	0.000E+00
450	0	0	0	1	1	1	5.095E+00	4.600E+00	3.897E-01	0.000E+00
451	0	0	0	1	1	1	5.698E+00	4.600E+00	3.897E-01	0.000E+00
452	0	0	0	1	1	1	6.433E+00	4.600E+00	3.897E-01	0.000E+00
453	0	0	0	1	1	1	6.859E+00	4.600E+00	3.897E-01	0.000E+00
454	0	0	0	1	1	1	7.285E+00	4.600E+00	3.897E-01	0.000E+00
455	0	0	0	1	1	1	7.711E+00	4.600E+00	3.897E-01	0.000E+00
456	0	0	0	1	1	1	8.139E+00	4.600E+00	3.897E-01	0.000E+00
457	0	0	0	1	1	1	3.233E+00	5.659E+00	3.882E-01	0.000E+00
458	0	0	0	1	1	1	3.556E+00	5.659E+00	3.882E-01	0.000E+00

459	1	1	1	1	1	1	3.306E+00	2.582E+00	4.364E-01	0.000E+00
460	1	1	1	1	1	1	3.556E+00	2.582E+00	4.364E-01	0.000E+00
461	1	1	1	1	1	1	3.306E+00	5.839E+00	4.364E-01	0.000E+00
462	1	1	1	1	1	1	3.556E+00	5.839E+00	4.364E-01	0.000E+00
463	0	0	0	1	1	1	4.460E+00	3.728E+00	4.820E-01	0.000E+00
464	0	0	0	1	1	1	4.460E+00	4.692E+00	4.820E-01	0.000E+00
465	0	0	0	1	1	1	6.859E+00	3.723E+00	4.872E-01	0.000E+00
466	0	0	0	1	1	1	7.285E+00	3.723E+00	4.872E-01	0.000E+00
467	0	0	0	1	1	1	7.711E+00	3.723E+00	4.872E-01	0.000E+00
468	0	0	0	1	1	1	8.139E+00	3.723E+00	4.872E-01	0.000E+00
469	0	0	0	1	1	1	6.859E+00	4.697E+00	4.872E-01	0.000E+00
470	0	0	0	1	1	1	7.285E+00	4.697E+00	4.872E-01	0.000E+00
471	0	0	0	1	1	1	7.711E+00	4.697E+00	4.872E-01	0.000E+00
472	0	0	0	1	1	1	8.139E+00	4.697E+00	4.872E-01	0.000E+00
473	0	0	0	1	1	1	6.433E+00	3.697E+00	5.127E-01	0.000E+00
474	0	0	0	1	1	1	6.433E+00	4.723E+00	5.127E-01	0.000E+00
475	0	0	0	1	1	1	4.347E+00	3.682E+00	5.283E-01	0.000E+00
476	0	0	0	1	1	1	4.347E+00	4.738E+00	5.283E-01	0.000E+00
477	0	0	0	1	1	1	4.686E+00	4.068E+00	5.324E-01	0.000E+00
478	0	0	0	1	1	1	4.970E+00	4.068E+00	5.324E-01	0.000E+00
479	0	0	0	1	1	1	5.095E+00	4.068E+00	5.324E-01	0.000E+00
480	0	0	0	1	1	1	5.698E+00	4.068E+00	5.324E-01	0.000E+00
481	0	0	0	1	1	1	6.433E+00	4.068E+00	5.324E-01	0.000E+00
482	0	0	0	1	1	1	6.859E+00	4.068E+00	5.324E-01	0.000E+00
483	0	0	0	1	1	1	7.285E+00	4.068E+00	5.324E-01	0.000E+00
484	0	0	0	1	1	1	7.711E+00	4.068E+00	5.324E-01	0.000E+00
485	0	0	0	1	1	1	8.139E+00	4.068E+00	5.324E-01	0.000E+00
486	0	0	0	1	1	1	4.686E+00	4.353E+00	5.324E-01	0.000E+00
487	0	0	0	1	1	1	4.970E+00	4.353E+00	5.324E-01	0.000E+00
488	0	0	0	1	1	1	5.095E+00	4.353E+00	5.324E-01	0.000E+00
489	0	0	0	1	1	1	5.698E+00	4.353E+00	5.324E-01	0.000E+00
490	0	0	0	1	1	1	6.433E+00	4.353E+00	5.324E-01	0.000E+00
491	0	0	0	1	1	1	6.859E+00	4.353E+00	5.324E-01	0.000E+00
492	0	0	0	1	1	1	7.285E+00	4.353E+00	5.324E-01	0.000E+00
493	0	0	0	1	1	1	7.711E+00	4.353E+00	5.324E-01	0.000E+00
494	0	0	0	1	1	1	8.139E+00	4.353E+00	5.324E-01	0.000E+00
495	0	0	0	1	1	1	5.095E+00	3.653E+00	5.568E-01	0.000E+00
496	0	0	0	1	1	1	5.698E+00	3.653E+00	5.568E-01	0.000E+00
497	0	0	0	1	1	1	5.095E+00	4.767E+00	5.568E-01	0.000E+00
498	0	0	0	1	1	1	5.698E+00	4.767E+00	5.568E-01	0.000E+00
499	0	0	0	1	1	1	3.233E+00	3.636E+00	5.745E-01	0.000E+00
500	0	0	0	1	1	1	3.556E+00	3.636E+00	5.745E-01	0.000E+00
501	0	0	0	1	1	1	3.712E+00	3.636E+00	5.745E-01	0.000E+00
502	0	0	0	1	1	1	3.837E+00	3.636E+00	5.745E-01	0.000E+00
503	0	0	0	1	1	1	3.962E+00	3.636E+00	5.745E-01	0.000E+00
504	0	0	0	1	1	1	4.030E+00	3.636E+00	5.745E-01	0.000E+00
505	0	0	0	1	1	1	4.098E+00	3.636E+00	5.745E-01	0.000E+00
506	0	0	0	1	1	1	4.166E+00	3.636E+00	5.745E-01	0.000E+00
507	0	0	0	1	1	1	4.233E+00	3.636E+00	5.745E-01	0.000E+00
508	0	0	0	1	1	1	3.233E+00	4.785E+00	5.745E-01	0.000E+00

509	0	0	0	1	1	1	3.556E+00	4.785E+00	5.745E-01	0.000E+00
510	0	0	0	1	1	1	3.712E+00	4.785E+00	5.745E-01	0.000E+00
511	0	0	0	1	1	1	3.837E+00	4.785E+00	5.745E-01	0.000E+00
512	0	0	0	1	1	1	3.962E+00	4.785E+00	5.745E-01	0.000E+00
513	0	0	0	1	1	1	4.030E+00	4.785E+00	5.745E-01	0.000E+00
514	0	0	0	1	1	1	4.098E+00	4.785E+00	5.745E-01	0.000E+00
515	0	0	0	1	1	1	4.166E+00	4.785E+00	5.745E-01	0.000E+00
516	0	0	0	1	1	1	4.233E+00	4.785E+00	5.745E-01	0.000E+00
517	1	1	1	1	1	1	3.306E+00	1.916E+00	6.147E-01	0.000E+00
518	1	1	1	1	1	1	3.556E+00	1.916E+00	6.147E-01	0.000E+00
519	1	1	1	1	1	1	3.306E+00	6.504E+00	6.147E-01	0.000E+00
520	1	1	1	1	1	1	3.556E+00	6.504E+00	6.147E-01	0.000E+00
521	0	0	0	1	1	1	4.970E+00	3.565E+00	6.452E-01	0.000E+00
522	0	0	0	1	1	1	4.970E+00	4.855E+00	6.452E-01	0.000E+00
523	0	0	0	1	1	1	4.460E+00	4.034E+00	6.584E-01	0.000E+00
524	0	0	0	1	1	1	4.460E+00	4.387E+00	6.584E-01	0.000E+00
525	0	0	0	1	1	1	3.962E+00	3.546E+00	6.638E-01	0.000E+00
526	0	0	0	1	1	1	4.030E+00	3.546E+00	6.638E-01	0.000E+00
527	0	0	0	1	1	1	4.098E+00	3.546E+00	6.638E-01	0.000E+00
528	0	0	0	1	1	1	4.166E+00	3.546E+00	6.638E-01	0.000E+00
529	0	0	0	1	1	1	4.233E+00	3.546E+00	6.638E-01	0.000E+00
530	0	0	0	1	1	1	6.859E+00	4.032E+00	6.655E-01	0.000E+00
531	0	0	0	1	1	1	7.285E+00	4.032E+00	6.655E-01	0.000E+00
532	0	0	0	1	1	1	7.711E+00	4.032E+00	6.655E-01	0.000E+00
533	0	0	0	1	1	1	8.139E+00	4.032E+00	6.655E-01	0.000E+00
534	0	0	0	1	1	1	6.859E+00	4.389E+00	6.655E-01	0.000E+00
535	0	0	0	1	1	1	7.285E+00	4.389E+00	6.655E-01	0.000E+00
536	0	0	0	1	1	1	7.711E+00	4.389E+00	6.655E-01	0.000E+00
537	0	0	0	1	1	1	8.139E+00	4.389E+00	6.655E-01	0.000E+00
538	0	0	0	1	1	1	3.962E+00	4.874E+00	6.638E-01	0.000E+00
539	0	0	0	1	1	1	4.030E+00	4.874E+00	6.638E-01	0.000E+00
540	0	0	0	1	1	1	4.098E+00	4.874E+00	6.638E-01	0.000E+00
541	0	0	0	1	1	1	4.166E+00	4.874E+00	6.638E-01	0.000E+00
542	0	0	0	1	1	1	4.233E+00	4.874E+00	6.638E-01	0.000E+00
543	0	0	0	1	1	1	6.433E+00	4.023E+00	7.004E-01	0.000E+00
544	0	0	0	1	1	1	6.433E+00	4.398E+00	7.004E-01	0.000E+00
545	0	0	0	1	1	1	4.347E+00	4.017E+00	7.216E-01	0.000E+00
546	0	0	0	1	1	1	4.347E+00	4.404E+00	7.216E-01	0.000E+00
547	0	0	0	1	1	1	3.233E+00	3.457E+00	7.531E-01	0.000E+00
548	0	0	0	1	1	1	3.556E+00	3.457E+00	7.531E-01	0.000E+00
549	0	0	0	1	1	1	3.712E+00	3.457E+00	7.531E-01	0.000E+00
550	0	0	0	1	1	1	3.837E+00	3.457E+00	7.531E-01	0.000E+00
551	0	0	0	1	1	1	3.962E+00	3.457E+00	7.531E-01	0.000E+00
552	0	0	0	1	1	1	4.030E+00	3.457E+00	7.531E-01	0.000E+00
553	0	0	0	1	1	1	4.098E+00	3.457E+00	7.531E-01	0.000E+00
554	0	0	0	1	1	1	4.166E+00	3.457E+00	7.531E-01	0.000E+00
555	0	0	0	1	1	1	4.233E+00	3.457E+00	7.531E-01	0.000E+00
556	0	0	0	1	1	1	4.347E+00	3.457E+00	7.531E-01	0.000E+00
557	0	0	0	1	1	1	4.460E+00	3.457E+00	7.531E-01	0.000E+00
558	0	0	0	1	1	1	4.686E+00	3.457E+00	7.531E-01	0.000E+00
559	0	0	0	1	1	1	4.970E+00	3.457E+00	7.531E-01	0.000E+00

560	0	0	0	1	1	1	3.233E+00	4.963E+00	7.531E-01	0.000E+00
561	0	0	0	1	1	1	3.556E+00	4.963E+00	7.531E-01	0.000E+00
562	0	0	0	1	1	1	3.712E+00	4.963E+00	7.531E-01	0.000E+00
563	0	0	0	1	1	1	3.837E+00	4.963E+00	7.531E-01	0.000E+00
564	0	0	0	1	1	1	3.962E+00	4.963E+00	7.531E-01	0.000E+00
565	0	0	0	1	1	1	4.030E+00	4.963E+00	7.531E-01	0.000E+00
566	0	0	0	1	1	1	4.098E+00	4.963E+00	7.531E-01	0.000E+00
567	0	0	0	1	1	1	4.166E+00	4.963E+00	7.531E-01	0.000E+00
568	0	0	0	1	1	1	4.233E+00	4.963E+00	7.531E-01	0.000E+00
569	0	0	0	1	1	1	4.347E+00	4.963E+00	7.531E-01	0.000E+00
570	0	0	0	1	1	1	4.460E+00	4.963E+00	7.531E-01	0.000E+00
571	0	0	0	1	1	1	4.686E+00	4.963E+00	7.531E-01	0.000E+00
572	0	0	0	1	1	1	4.970E+00	4.963E+00	7.531E-01	0.000E+00
573	0	0	0	1	1	1	5.095E+00	4.006E+00	7.606E-01	0.000E+00
574	0	0	0	1	1	1	5.698E+00	4.006E+00	7.606E-01	0.000E+00
575	0	0	0	1	1	1	5.095E+00	4.414E+00	7.606E-01	0.000E+00
576	0	0	0	1	1	1	5.698E+00	4.414E+00	7.606E-01	0.000E+00
577	0	0	0	1	1	1	3.233E+00	4.000E+00	7.848E-01	0.000E+00
578	0	0	0	1	1	1	3.556E+00	4.000E+00	7.848E-01	0.000E+00
579	0	0	0	1	1	1	3.712E+00	4.000E+00	7.848E-01	0.000E+00
580	0	0	0	1	1	1	3.837E+00	4.000E+00	7.848E-01	0.000E+00
581	0	0	0	1	1	1	3.962E+00	4.000E+00	7.848E-01	0.000E+00
582	0	0	0	1	1	1	4.030E+00	4.000E+00	7.848E-01	0.000E+00
583	0	0	0	1	1	1	4.098E+00	4.000E+00	7.848E-01	0.000E+00
584	0	0	0	1	1	1	4.166E+00	4.000E+00	7.848E-01	0.000E+00
585	0	0	0	1	1	1	4.233E+00	4.000E+00	7.848E-01	0.000E+00
586	0	0	0	1	1	1	3.233E+00	4.420E+00	7.848E-01	0.000E+00
587	0	0	0	1	1	1	3.556E+00	4.420E+00	7.848E-01	0.000E+00
588	0	0	0	1	1	1	3.712E+00	4.420E+00	7.848E-01	0.000E+00
589	0	0	0	1	1	1	3.837E+00	4.420E+00	7.848E-01	0.000E+00
590	0	0	0	1	1	1	3.962E+00	4.420E+00	7.848E-01	0.000E+00
591	0	0	0	1	1	1	4.030E+00	4.420E+00	7.848E-01	0.000E+00
592	0	0	0	1	1	1	4.098E+00	4.420E+00	7.848E-01	0.000E+00
593	0	0	0	1	1	1	4.166E+00	4.420E+00	7.848E-01	0.000E+00
594	0	0	0	1	1	1	4.233E+00	4.420E+00	7.848E-01	0.000E+00
595	0	0	0	1	1	1	3.712E+00	3.348E+00	8.627E-01	0.000E+00
596	0	0	0	1	1	1	3.712E+00	5.073E+00	8.627E-01	0.000E+00
597	0	0	0	1	1	1	4.970E+00	3.974E+00	8.813E-01	0.000E+00
598	0	0	0	1	1	1	4.970E+00	4.446E+00	8.813E-01	0.000E+00
599	0	0	0	1	1	1	3.962E+00	3.967E+00	9.068E-01	0.000E+00
600	0	0	0	1	1	1	4.030E+00	3.967E+00	9.068E-01	0.000E+00
601	0	0	0	1	1	1	4.098E+00	3.967E+00	9.068E-01	0.000E+00
602	0	0	0	1	1	1	4.166E+00	3.967E+00	9.068E-01	0.000E+00
603	0	0	0	1	1	1	4.233E+00	3.967E+00	9.068E-01	0.000E+00
604	0	0	0	1	1	1	3.962E+00	4.453E+00	9.068E-01	0.000E+00
605	0	0	0	1	1	1	4.030E+00	4.453E+00	9.068E-01	0.000E+00
606	0	0	0	1	1	1	4.098E+00	4.453E+00	9.068E-01	0.000E+00
607	0	0	0	1	1	1	4.166E+00	4.453E+00	9.068E-01	0.000E+00
608	0	0	0	1	1	1	4.233E+00	4.453E+00	9.068E-01	0.000E+00
609	0	0	0	1	1	1	3.233E+00	3.238E+00	9.723E-01	0.000E+00

610	0	0	0	1	1	1	3.556E+00	3.238E+00	9.723E-01	0.000E+00
611	0	0	0	1	1	1	3.712E+00	3.238E+00	9.723E-01	0.000E+00
612	0	0	0	1	1	1	3.233E+00	5.182E+00	9.723E-01	0.000E+00
613	0	0	0	1	1	1	3.556E+00	5.182E+00	9.723E-01	0.000E+00
614	0	0	0	1	1	1	3.712E+00	5.182E+00	9.723E-01	0.000E+00
615	0	0	0	1	1	1	3.233E+00	3.935E+00	1.029E+00	0.000E+00
616	0	0	0	1	1	1	3.556E+00	3.935E+00	1.029E+00	0.000E+00
617	0	0	0	1	1	1	3.712E+00	3.935E+00	1.029E+00	0.000E+00
618	0	0	0	1	1	1	3.837E+00	3.935E+00	1.029E+00	0.000E+00
619	0	0	0	1	1	1	3.962E+00	3.935E+00	1.029E+00	0.000E+00
620	0	0	0	1	1	1	4.030E+00	3.935E+00	1.029E+00	0.000E+00
621	0	0	0	1	1	1	4.098E+00	3.935E+00	1.029E+00	0.000E+00
622	0	0	0	1	1	1	4.166E+00	3.935E+00	1.029E+00	0.000E+00
623	0	0	0	1	1	1	4.233E+00	3.935E+00	1.029E+00	0.000E+00
624	0	0	0	1	1	1	4.347E+00	3.935E+00	1.029E+00	0.000E+00
625	0	0	0	1	1	1	4.460E+00	3.935E+00	1.029E+00	0.000E+00
626	0	0	0	1	1	1	4.686E+00	3.935E+00	1.029E+00	0.000E+00
627	0	0	0	1	1	1	4.970E+00	3.935E+00	1.029E+00	0.000E+00
628	0	0	0	1	1	1	3.233E+00	4.486E+00	1.029E+00	0.000E+00
629	0	0	0	1	1	1	3.556E+00	4.486E+00	1.029E+00	0.000E+00
630	0	0	0	1	1	1	3.712E+00	4.486E+00	1.029E+00	0.000E+00
631	0	0	0	1	1	1	3.837E+00	4.486E+00	1.029E+00	0.000E+00
632	0	0	0	1	1	1	3.962E+00	4.486E+00	1.029E+00	0.000E+00
633	0	0	0	1	1	1	4.030E+00	4.486E+00	1.029E+00	0.000E+00
634	0	0	0	1	1	1	4.098E+00	4.486E+00	1.029E+00	0.000E+00
635	0	0	0	1	1	1	4.166E+00	4.486E+00	1.029E+00	0.000E+00
636	0	0	0	1	1	1	4.233E+00	4.486E+00	1.029E+00	0.000E+00
637	0	0	0	1	1	1	4.347E+00	4.486E+00	1.029E+00	0.000E+00
638	0	0	0	1	1	1	4.460E+00	4.486E+00	1.029E+00	0.000E+00
639	0	0	0	1	1	1	4.686E+00	4.486E+00	1.029E+00	0.000E+00
640	0	0	0	1	1	1	4.970E+00	4.486E+00	1.029E+00	0.000E+00
641	0	0	0	1	1	1	3.233E+00	3.150E+00	1.061E+00	0.000E+00
642	0	0	0	1	1	1	3.556E+00	3.150E+00	1.061E+00	0.000E+00
643	0	0	0	1	1	1	3.233E+00	5.271E+00	1.061E+00	0.000E+00
644	0	0	0	1	1	1	3.556E+00	5.271E+00	1.061E+00	0.000E+00
645	0	0	0	1	1	1	3.712E+00	3.894E+00	1.178E+00	0.000E+00
646	0	0	0	1	1	1	3.712E+00	4.526E+00	1.178E+00	0.000E+00
647	1	1	1	1	1	1	3.306E+00	3.018E+00	1.192E+00	0.000E+00
648	1	1	1	1	1	1	3.556E+00	3.018E+00	1.192E+00	0.000E+00
649	1	1	1	1	1	1	3.306E+00	5.402E+00	1.192E+00	0.000E+00
650	1	1	1	1	1	1	3.556E+00	5.402E+00	1.192E+00	0.000E+00
651	0	0	0	1	1	1	3.233E+00	3.854E+00	1.328E+00	0.000E+00
652	0	0	0	1	1	1	3.556E+00	3.854E+00	1.328E+00	0.000E+00
653	0	0	0	1	1	1	3.712E+00	3.854E+00	1.328E+00	0.000E+00
654	0	0	0	1	1	1	3.233E+00	4.566E+00	1.328E+00	0.000E+00
655	0	0	0	1	1	1	3.556E+00	4.566E+00	1.328E+00	0.000E+00
656	0	0	0	1	1	1	3.712E+00	4.566E+00	1.328E+00	0.000E+00
657	0	0	0	1	1	1	3.233E+00	3.822E+00	1.449E+00	0.000E+00
658	0	0	0	1	1	1	3.556E+00	3.822E+00	1.449E+00	0.000E+00
659	0	0	0	1	1	1	3.233E+00	4.598E+00	1.449E+00	0.000E+00

```
660  0  0  0  1  1  1 3.556E+00 4.598E+00 1.449E+00 0.000E+00
661  1  1  1  1  1  1 3.306E+00 3.774E+00 1.629E+00 0.000E+00
662  1  1  1  1  1  1 3.556E+00 3.774E+00 1.629E+00 0.000E+00
663  1  1  1  1  1  1 3.306E+00 4.647E+00 1.629E+00 0.000E+00
664  1  1  1  1  1  1 3.556E+00 4.647E+00 1.629E+00 0.000E+00
665  1  1  1  1  1  1 3.306E+00 2.531E+00 1.679E+00 0.000E+00
666  1  1  1  1  1  1 3.556E+00 2.531E+00 1.679E+00 0.000E+00
667  1  1  1  1  1  1 3.306E+00 5.890E+00 1.679E+00 0.000E+00
668 .1  1  1  1  1  1 3.556E+00 5.890E+00 1.679E+00 0.000E+00
669  1  1  1  1  1  1 3.306E+00 3.596E+00 2.294E+00 0.000E+00
670  1  1  1  1  1  1 3.556E+00 3.596E+00 2.294E+00 0.000E+00
671  1  1  1  1  1  1 3.306E+00 4.825E+00 2.294E+00 0.000E+00
672  1  1  1  1  1  1 3.556E+00 4.825E+00 2.294E+00 0.000E+00
```
**** PRINT OF EQUATION NUMBERS SUPPRESSED

1**** 8-NODE BRICK ELEMENTS

number of elements   =   420
number of materials =   1
number of load types =   0

1**** MATERIAL DATA

INDEX	E	MU	WEIGHT DENSITY	ALPHA	SHEAR MODULUS
1	3.0000E+07	3.0000E-01	2.8360E-01	6.5000E-06	1.1538E+07

Figure 7.2 shows the deflection of the spindle and hub, and Fig. 7.3 shows the stress levels in the spindle and hub. Figure 7.4 shows the deflection contours for the spindle-hub combination.

The 4 mm diameter hole was added, and new stress values calculated. Again, the following listing gives the geometry for the model with the hole. This geometry is good for both cases of hole-load alignment.

Figure 7.5 is a close-up of the stress levels around the hole and Fig. 7.6 illustrates the stress levels for the entire spindle-hub combination with the hole aligned in the direction of the load.

Figure 7.7 shows the deflections while Fig. 7.8 shows the deflection contours on the spindle-hub combination.

Figures 7.9 through 7.12 show the results of the analysis for the hole aligned 90° to the loads.

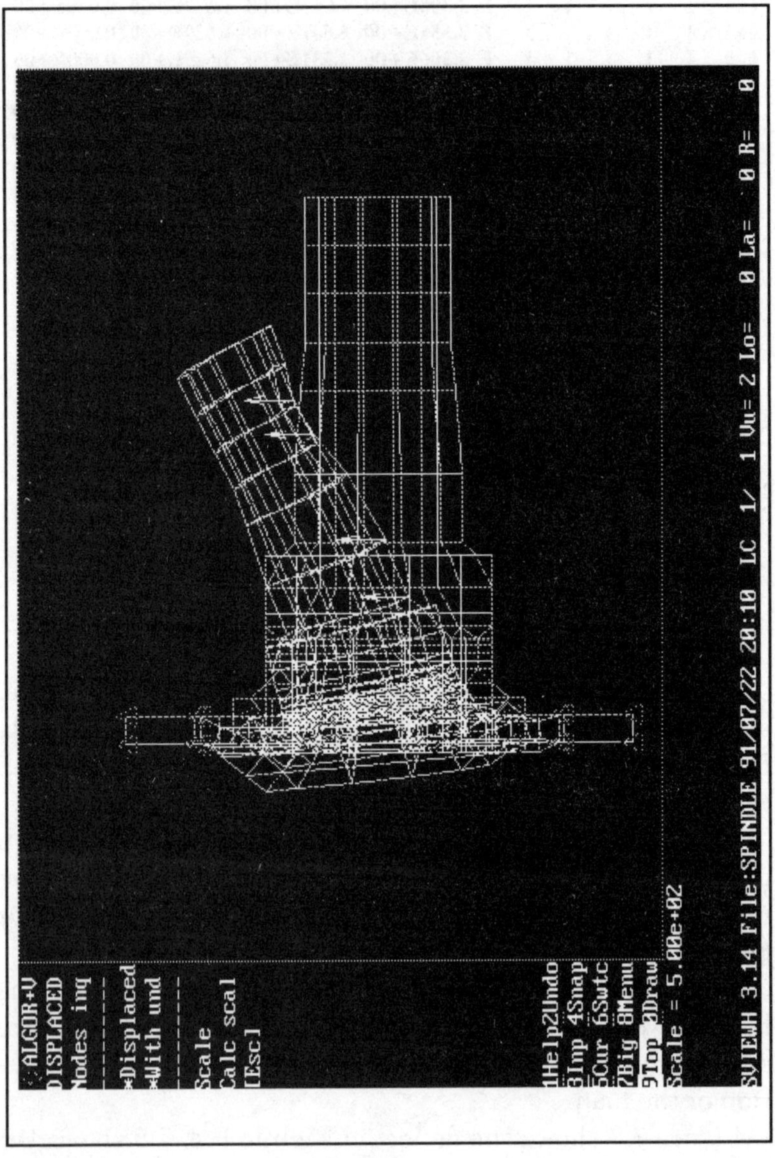

Figure 7.2 Deformed spindle baseline model.

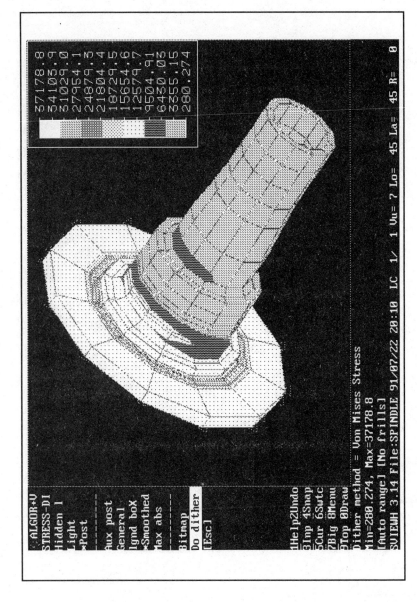

Figure 7.3 Stress levels in baseline model.

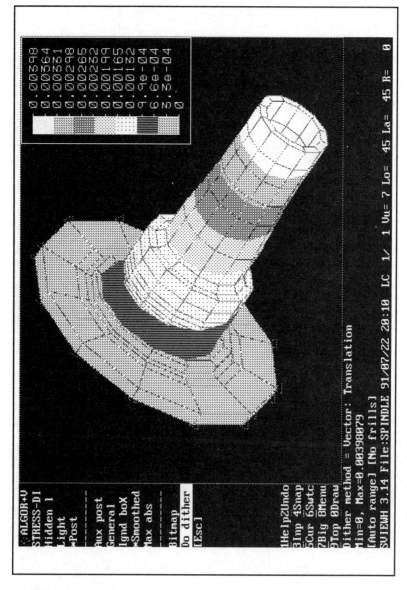

Figure 7.4 Spindle-hub deflection contours—no hole.

184

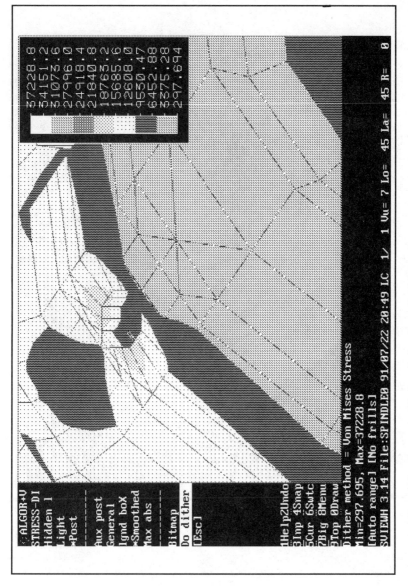

Figure 7.5 Stress levels around hole—hole aligned with load.

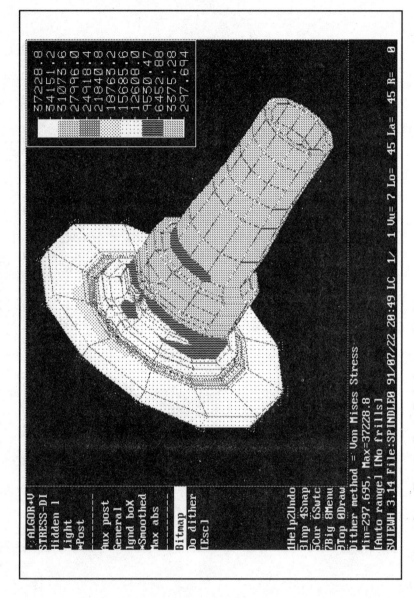

Figure 7.6 Overall view of spindle-hub stress levels.

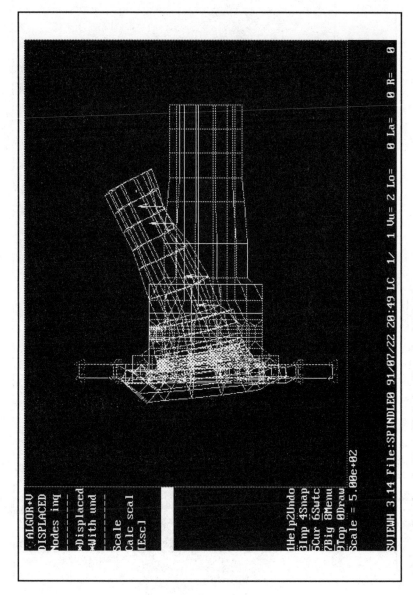

Figure 7.7 Displaced view with hole aligned with load.

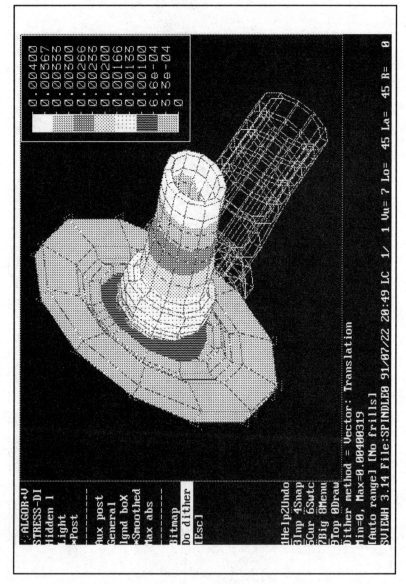

Figure 7.8 Displacement contours for hole aligned with load.

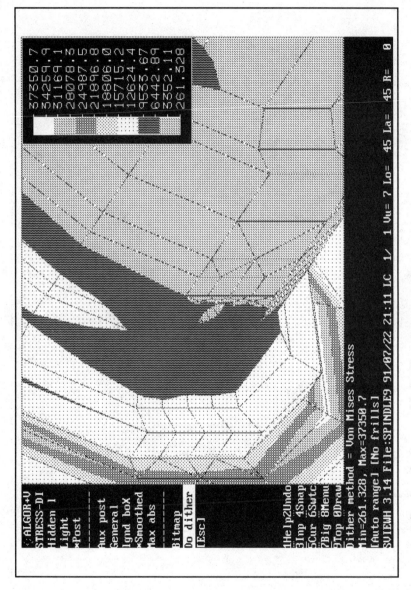

Figure 7.9 Stress values around hole aligned 90° to load.

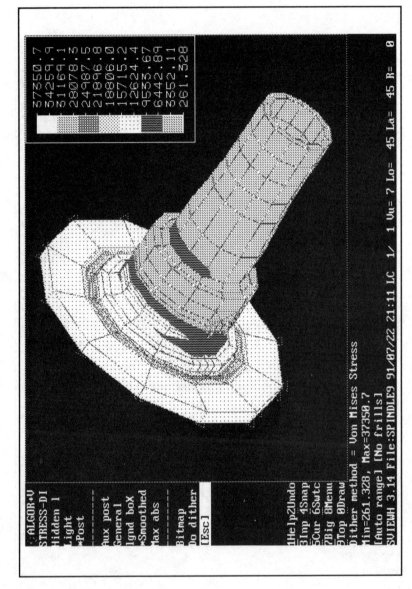

Figure 7.10 Stress values for spindle-hub—hole 90° to load.

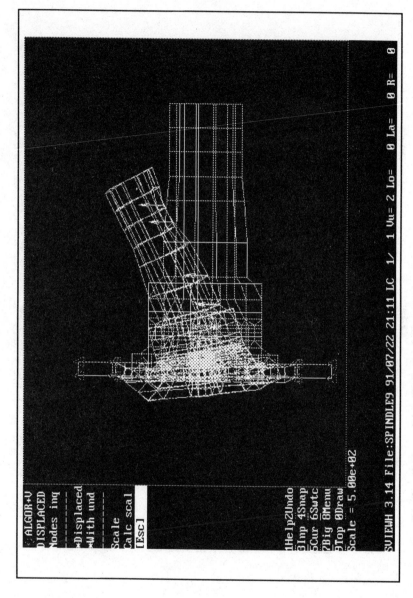

Figure 7.11 Displaced view of spindle-hub—hole 90° from load.

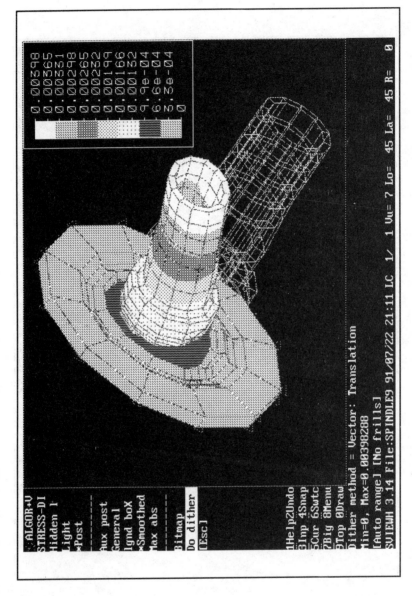

Figure 7.12 Displaced contours—hole 90° from load.

7.4   Results of Spindle-Hub Analysis

With the hole aligned with the loads, stress increased about 15 percent. With the hole at 90° to the load, no significant increase in stress was seen. Deflections among the three cases are consistent.

Mesh refinement was guided by looking at stress values at adjacent nodes. The mesh was refined to the point where adjacent nodes had a stress ratio of under 2:1. Due to the small size of the hole compared to the rest of the part, it was felt that the stress at the nodes was the best method for refining the mesh. This spindle-hub system is going into production in mid-1991. Neither of the vehicles for which this system is designed to fit are currently exported to the United States.

The same model was used to run a modal analysis on the spindle. If the fundamental frequencies of the spindle unit are near other resonant points in the suspension system, it is quite possible that the design should be modified to smooth out the suspension characteristics. Results of the modal analysis are given in Figs. 7.13 through 7.16. As can be seen, the first four fundamental frequencies are quite high and probably would not be noticeable in the overall suspension feel. The only problem that might arise would be in the additional tubing added to the spindle. If the section of tubing exposed were to have a natural frequency in the area of the frequencies shown, it would be appropriate to move the fundamental frequency of the small tube by means of a support.

7.5   Using FEA for Maximum Fatigue Life and Minimum Weight

Webb Wheel Products, Inc. of Cullman, Alabama (a member of the Marmon Group of companies)[29] uses FEA in the design process of bringing to market commercial vehicle wheel hubs for a wide variety of vehicles. Minimizing the prototype cycle for a new part was a goal of integrating FEA

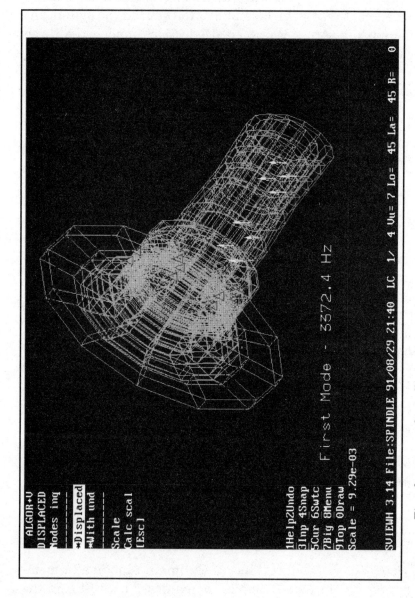

Figure 7.13 First frequency of spindle.

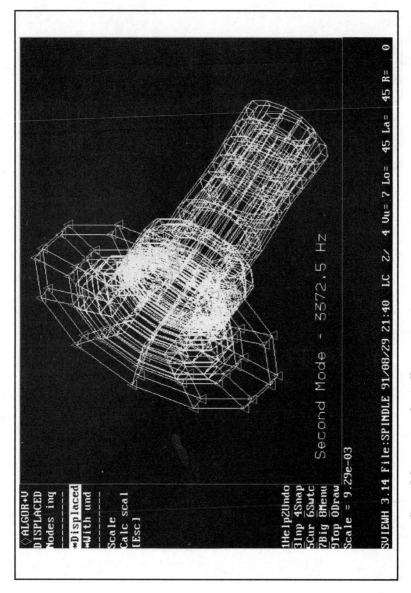

Figure 7.14 Second frequency of spindle.

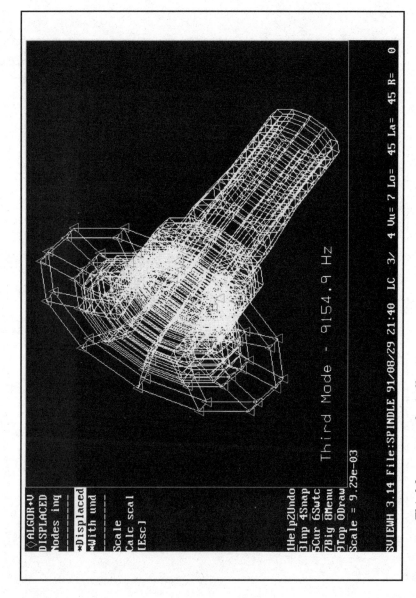

Figure 7.15 Third frequency of spindle.

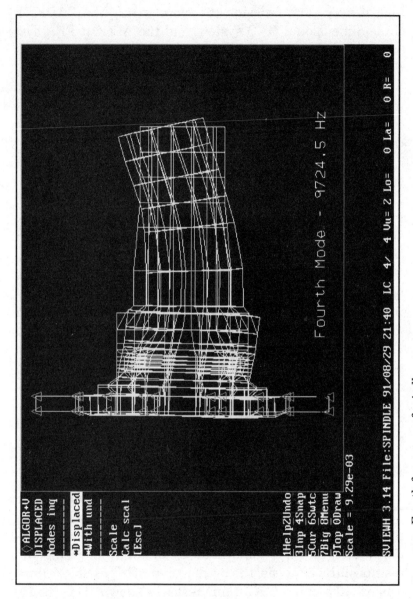

Figure 7.16 Fourth frequency of spindle.

in the overall design process. In general, it takes about six months and several thousand dollars to build a prototype and, if necessary, redesign the piece. Using the CADKEY three-dimensional CAD system and the ALGOR FEA system allows the company to bring the finished part to the field sooner than previously had been accomplished.

For the particular part used in this text, it was necessary to reduce product liability risks and to reduce the material requirements. Savings were substantial since this product carried a long production run.

The 45 lb hub was made from a cast steel grade almost equivalent to SAE 1030 steel and is shown in Figure 7.17. The hub has a bolt pattern of 22 mm studs to hold the wheel in place and 0.75-inch bosses for rotor mounting on the back side and 0.56-inch bosses for flange mounting on the front side.

The critical area of the hub was in the vicinity of the bosses. Using the ALGOR preprocessor programs and transferring data from the CAD package allowed rapid model creation. Figure 7.18 shows the CAD profile drawing. Figure 7.19 shows a three-dimensional section of the hub while Fig. 7.20 shows the one-fourth section used for the model rendered as a solid.

The model was run through the linear static processor to determine the deflection and stress results. Figs. 7.21 and 7.22 show the surface stresses on the outside and inside of the hub, respectively.

For comparison, the hub was restrained and loaded in the same manner as that prescribed in the SAE fatigue test for hubs. The SAE test is a rotary bending fatigue test based on the offset of the wheel, the axle loading, a 1.4 test acceleration factor, and a bending load of 13,481 ft·lb. Using the results of the analysis, the predicted area of possible failure was exactly where the fatigue test had shown. The results also indicated a stress level where the safe million-cycle fatigue limit would occur. This analysis allowed minor

Figure 7.17   Photograph of hub after FEA and build.

199

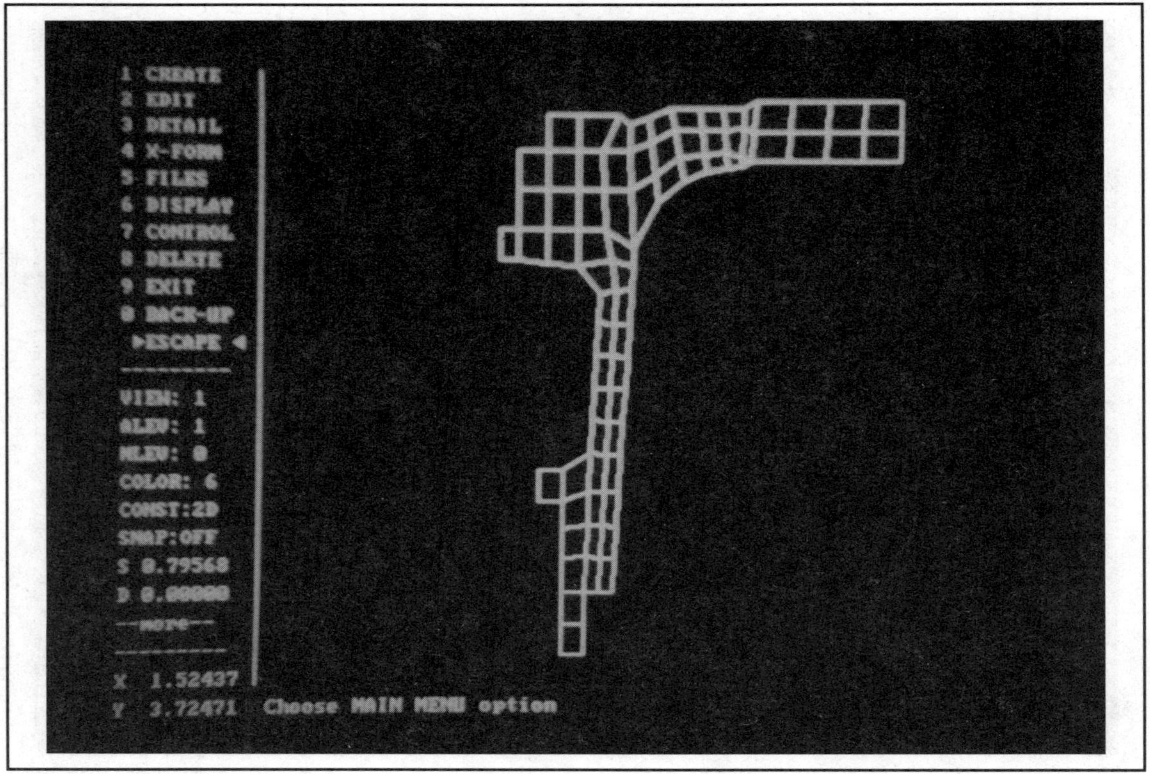

Figure 7.18 Hub profile.

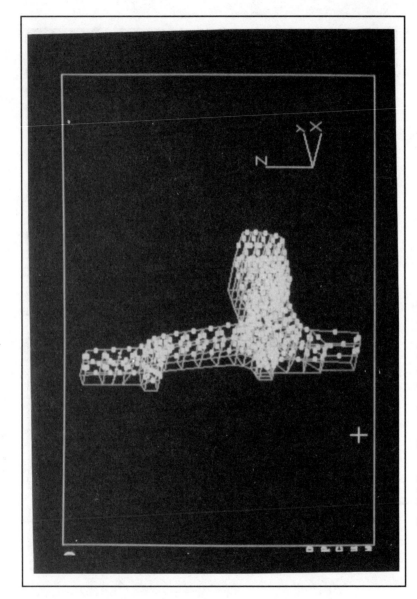

Figure 7.19 Partial three-dimensional model.

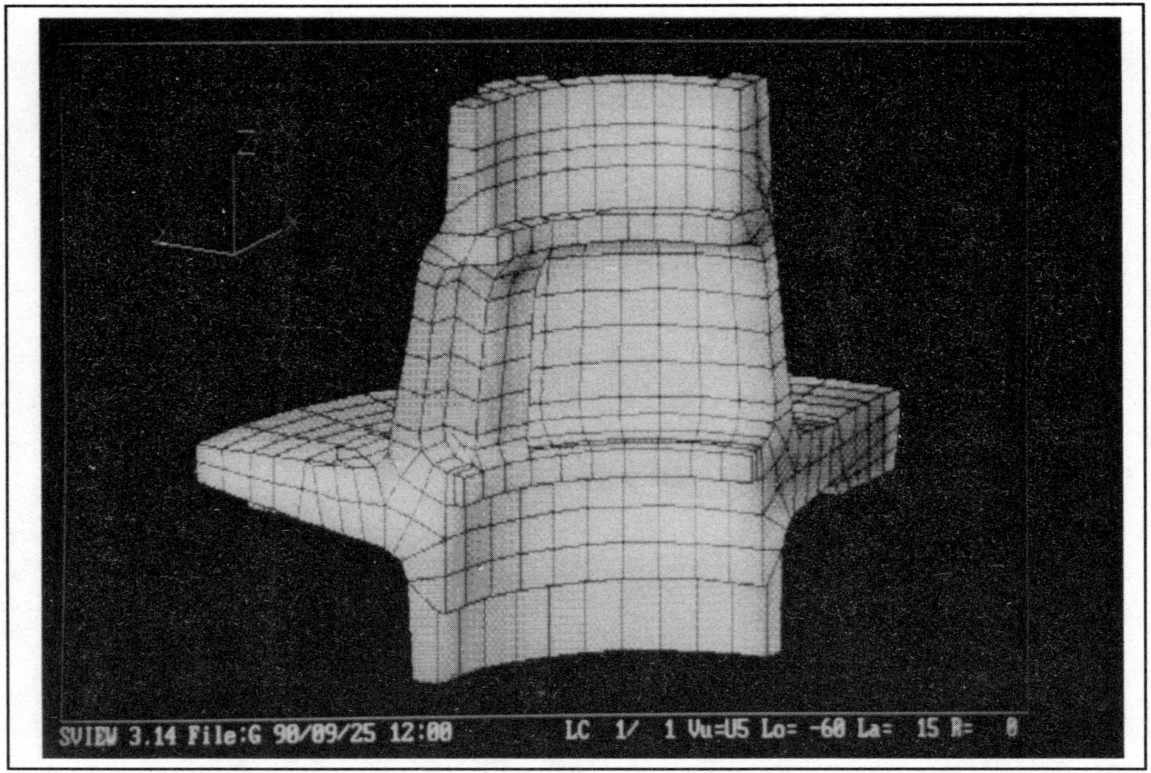

SVIEW 3.14 File:G 90/09/25 12:00   LC 1/ 1 Vw=05 Lo= -60 La= 15 R= 0

Figure 7.20 One-fourth hub solid model.

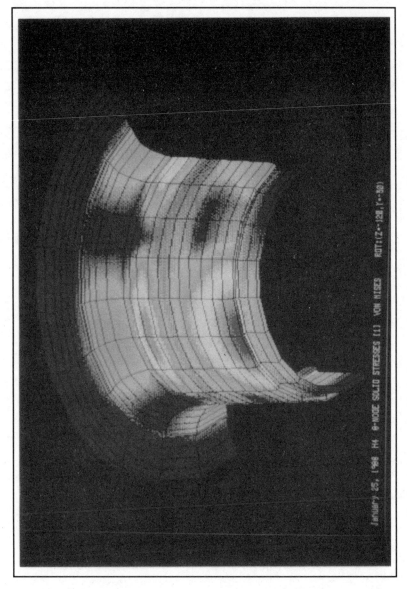

Figure 7.21 Surface stresses inside hub.

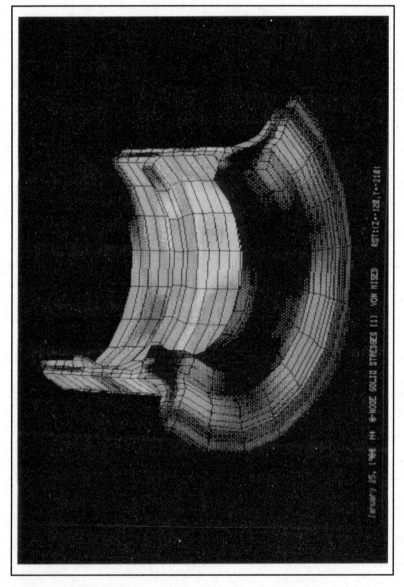

Figure 7.22 Surface stresses outside hub.

204

design modifications to meet the million-cycle design requirement.

7.6   Vehicle Aerodynamic Studies

One of the strong selling features of any product is the appearance of that product. Automotive designers must take into account the appearance and the functionality of the automobile that they are trying to sell to the public. Engineers in the automobile industry can often streamline vehicle designs by using flow analysis to predict flow separation points on a body. The next example in this chapter is the determination of flow and pressure distribution using the Fluids package from ALGOR. Based on the Penalty method, SuperFlow solves for the fluid velocity field and the pressure field. The examples given in this chapter are performed using the two-dimensional incompressible viscous flow steady state version of the software package.

Simulation of flow fields has become extremely important for understanding complex flows where no closed solution exists. Fluid flow can be classified as laminar or turbulent, depending on a number of factors, and they can be classified as compressible or incompressible, depending on the Mach number, and viscid or inviscid depending on the fluid idealization. Although the governing equations for incompressible flow are the conservation of momentum and the conservation of mass, inherent nonlinearity and a high Reynolds number lead to difficulties in the numerical solution of the equations.

Typical applications for incompressible flow simulation include:

Contained flows such as heated cavities

Flow past obstacles such as airfoils, vehicles, etc.

Flow through pipes, ducts of various configurations

Flow in heat exchangers

Flow in lubrication bearings, etc.

The incompressible Navier-Stokes equations are the momentum equations subject to incompressibility constraints. In addition, a simple linear constitutive relation between stress and the rate of strain is assumed. The basic equations are given below.

$$\rho\left(\frac{\partial u}{\partial t} + u\nabla u\right) + \nabla p - \mu\nabla^2 u = 0$$

$$\nabla u = 0$$

(7.1)

where u represents the velocity, p is the pressure, ρ is the density, and μ is the viscosity. The above equations are the velocity-pressure (primitive) formulation. There are two unknown velocity components and one pressure component for a two-dimensional problem. The advantages of this method are that the physical quantities u_1, u_2, and p are computed directly, and boundary conditions are easy to implement. Also, this method is equally applicable in two and three dimensions.

Everything has its penalties. For a three-dimensional flow, the number of unknowns is four. It is not possible to use equal orders of interpolation for the velocity and pressure and there may be spurious pressure nodes. Also the pressure field and secondary variables like vorticity and stream function need to be recovered.

For two-dimensional flows, the basic variables can be transformed into a set of equations containing two variables, the stream function and the vorticity variable. The ALGOR program uses the velocity-pressure formulation.

Any effort directed at the numerical simulation of the velocity-pressure formulation requires the discretization of the governing equations. The difficulties associated with the finite element modeling of the governing equations described above include the nonlinearity due to the convection term and the incompressibility constraint. It has been noted that not all combinations of velocity-pressure interpolations yield a stable solution scheme. There are two techniques that are popular for solving the equations. These are the mixed methods that solve for both velocity and pressure, and the penalty method. In the penalty method the pressure is eliminated by using the slightly compressible form of the incompressible constraint. This is the method used in the ALGOR programs.

The incompressibility constraint, the second part of Eq. 7.1 is eliminated using a penalty parameter. Therefore, only the velocity components are computed. Once the velocities are calculated, the pressure field is evaluated by the following formula

$$p = -\lambda \nabla u \qquad (7.2)$$

where λ is the penalty parameter.

The pressure field is determined to within an arbitrary constant from the above formula. To obtain the actual pressures when using the data, a pressure value within the computational domain must be known in order to relate the computed pressure field to real-world values.

7.7   Mini Van Model

A simple profile of a small van is given in Fig. 7.23. The purpose of the analysis is to see how profile changes for this two-dimensional model affect the velocity, pressure, and vorticity profiles. The analysis package used was the ALGOR SuperFlow routine. A van model was created using the SuperDraw II software. Boundary conditions were applied of zero velocity at the base (a wall), zero velocity at the van exterior, $v_z = 0$ at the top of the model as a plane of symmetry, and an inlet velocity of one to the calculational mesh. All values are dimensionless and all calculated values of velocity, pressure, and vorticity are dimensionless.

Figure 7.24 shows the velocity vector plot followed by the velocity contour plot given in Fig. 7.25.

Figure 7.26 illustrates the pressure distribution followed

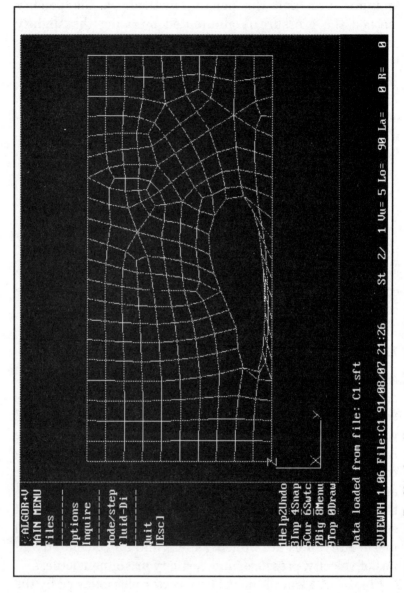

Figure 7.23 Profile of small van model with FEA mesh.

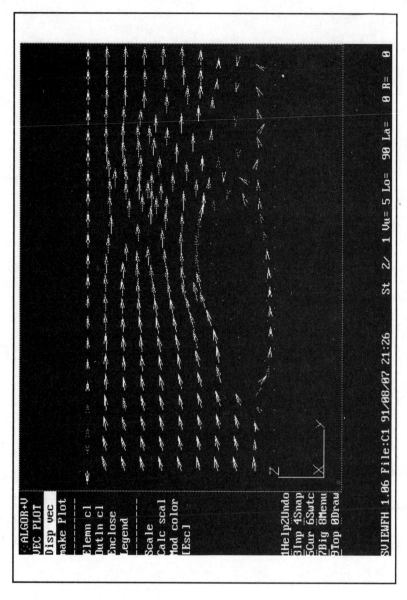

Figure 7.24 Van model—velocity vector plot.

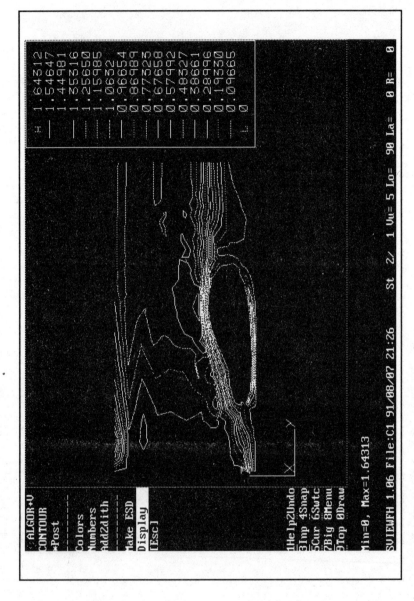

Figure 7.25 Van model—velocity contour plot.

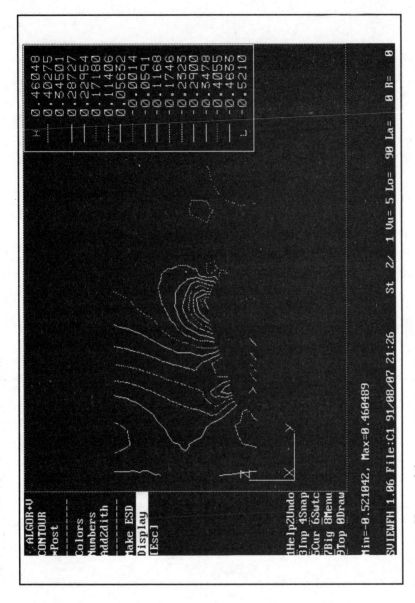

Figure 7.26 Van model—pressure contours.

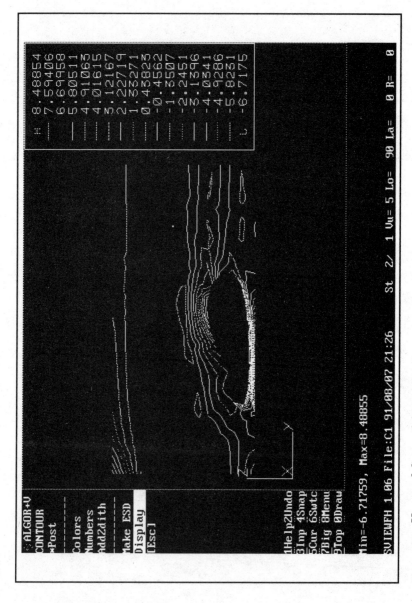

Figure 7.27 Van model—vorticity contours.

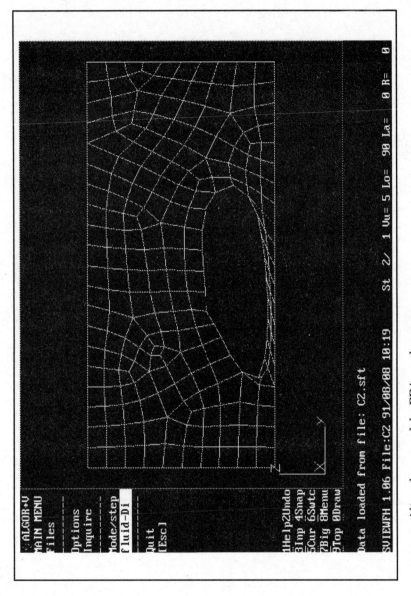

Figure 7.28 Altered van model—FEA mesh.

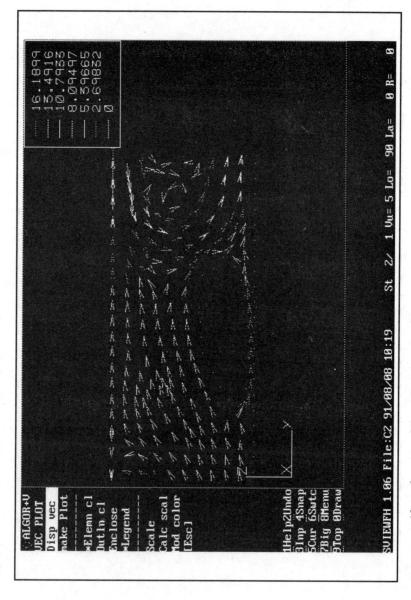

Figure 7.29 Altered van model—velocity vector plot.

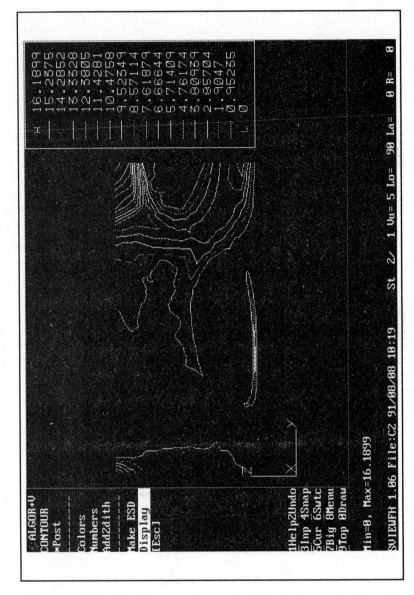

Figure 7.30 Altered van model—velocity contour plot.

215

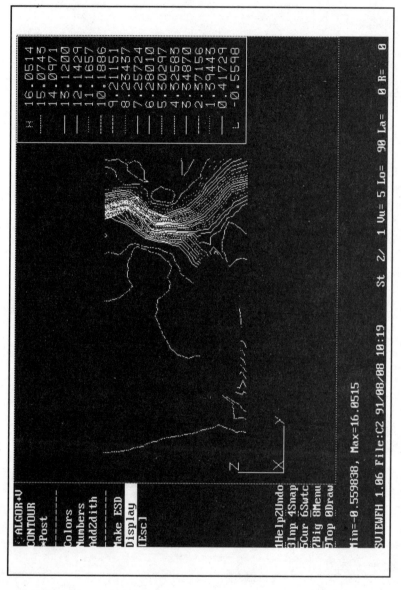

Figure 7.31 Altered van model—pressure contours.

216

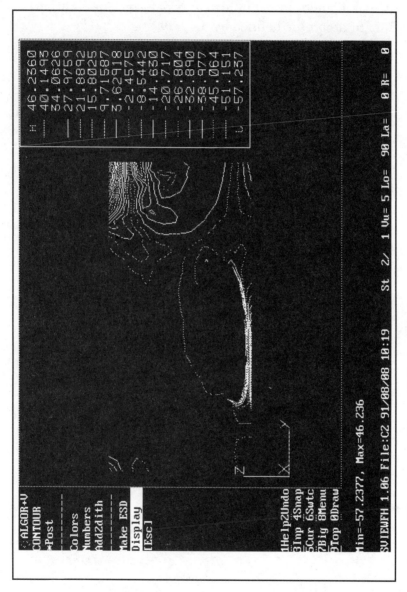

Figure 7.32 Altered van model—vorticity contours.

by the vorticity plot in Fig. 7.27. Recall that vorticity deals with the rotation of a fluid element. A fluid not only may translate and rotate, but also may deform. The rotation of a fluid at a point is defined by

$$\omega = 0.5 \nabla \times q \qquad (7.3)$$

When $\omega = 0$ throughout certain portions of a fluid, the motion is described as irrotational. The vorticity vector has certain characteristics similar to the velocity vector. Vortex lines are everywhere tangent to the vorticity vector.

The next series of figures illustrates the changes made in altering the profile of the original proposed small van vehicle. Figure 7.28 shows the body profile and computational grid. All boundary conditions are the same for this version as the previous version.

Again, the velocity profiles, pressure distribution, and vorticity plots are shown in Figs. 7.29 through 7.32.

8

Music Products and FEA

8.1 Introduction

As products become more sophisticated and the demands of consumers increase, the application of finite element analysis in product development and aftermarket product refinement increase. This chapter looks at an aftermarket application as applied to instrumental mouthpiece design. We will examine a possible improvement to a trumpet and, in particular, to the trumpet mouthpiece.

Most musicians look to gain a performance edge through long, dedicated hours of practice, a new instrument, modifications to their basic instrument, or a combination of the three. Certainly no one can discount the merits of practice. However, there are things that can be done to the instrument to make playing easier and/or to enhance the instrument characteristics.

After the musician settles on a particular brand of instrument, modifications to the basic feel of the instrument can be made.

8.2 The Trumpet Mouthpiece Problem

The analysis in the following sections addresses the modifications of a trumpet mouthpiece. The geometry is based on a standard Schilke 14A4 trumpet mouthpiece. Initially, the first four fundamental vibration modes are determined; then an attempt is made to modify the vibration characteristics. This modification can produce a more centered tone over the musical range of the trumpet by reducing or damping some vibration natural frequencies in the mouthpiece. When the lips buzz to produce the sound, they will vibrate at a certain pitch. For example, the note A in the staff (for treble clef instruments) has a frequency of 440 Hz. As you approach the natural frequency of the mouthpiece, the forcing function (your vibrating lips) excites the mouthpiece. The once clean vibrations that are carried through the trumpet are now amplified and carry into the horn the overtones of that note and the tone that is modified by the vibrating mouthpiece.

The purpose of what is known as a tone intensifier ring is to damp out the unwanted overtones and other frequencies so that the horn reproduces exactly what is intended by the player. This means that the natural frequencies of the mouthpiece are changed to a higher frequency to avoid hitting the resonant modes of the mouthpiece. A tone intensifier ring for this study is essentially a brass sleeve that fits over a portion of the mouthpiece. Usually, the ring will extend from the leadpipe/mouthpiece junction to about one inch up the mouthpiece shank toward the player.

It is open to question whether or not the tone intensifier ring actually works. Many players use and swear by the intensifier. My use of these devices indicates that they do change the sound and give a more centered sound. Companies such as The Selmer Company of Elkhart, Indiana are offering thicker-walled mouthpieces in addition to their regular line of mouthpieces. After the performance of the tone intensifier ring is examined, a sample case for a thicker-walled mouthpiece is run for comparison.

To model the mouthpiece, a somewhat common mouthpiece was used. All data were taken from published specifications and verified by caliper measurements.

8.2.1   Building the model

The basic model was constructed in SuperDraw II using lines and arcs. This is shown in Fig. 8.1. For all other models, this profile was used.

Using the *Generate* option, a mesh was constructed within the outline of that shown in Fig. 8.1. Also added at this stage were the restraints at the end of the mouthpiece. The goal is to do as much as possible to the profile that will be carried over to the full model without having to add anything to the full model later. Figure 8.2 illustrates the meshed profile with restraints before making the full model.

This profile was then rotated about the y-axis in intervals of 30° for twelve sections. For the rotation all restraints were placed around the model as shown in Fig. 8.3. Note that for problems of this type, it is easier to put any restraints and/or forces on the model during the model construction stage. As the full model is developed, the program automatically resolves forces and restraints relative to their position in the global field.

The model is next preprocessed by the ALGOR decoder into a format suitable for the processor—in this case, the Linear Dynamic Processor, SSAP1H. It is at this point that the material properties are inserted (for this case, brass) and the elements are defined (eight-node brick elements). During the decoding, a file is created that allows viewing of the solid model as well as deflections and stress levels. This stage of the analysis gives the engineer the chance to view the model from any perspective in order to see if any feature or element is missing, and if the restraints and/or forces are in the appropriate places. This is a handy and necessary feature of the ALGOR visualization package. This (ease of viewing the model) should be a requirement for any finite element program.

8.2.2   Results of the analysis

The model was analyzed using the Linear Dynamic processor to determine the natural frequencies up to the third

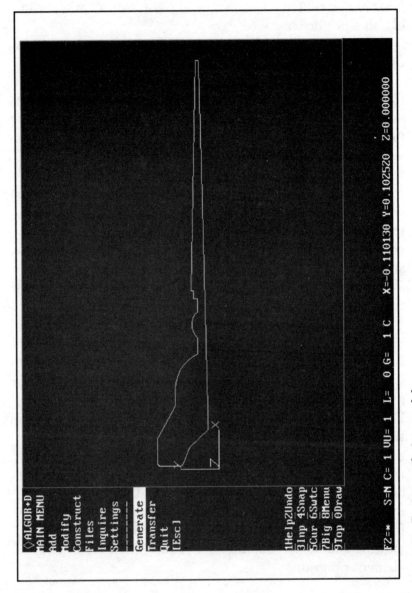

Figure 8.1 Basic mouthpiece model.

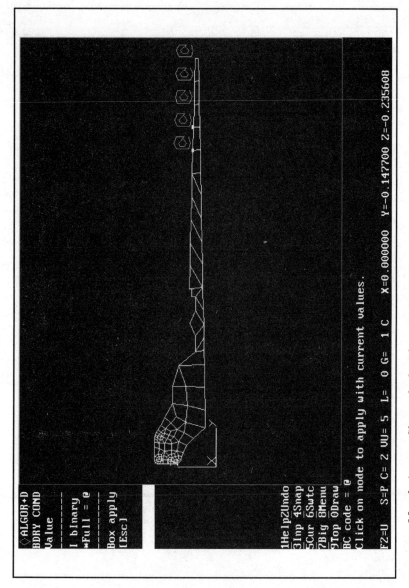

Figure 8.2 Mouthpiece profile meshed with restraints.

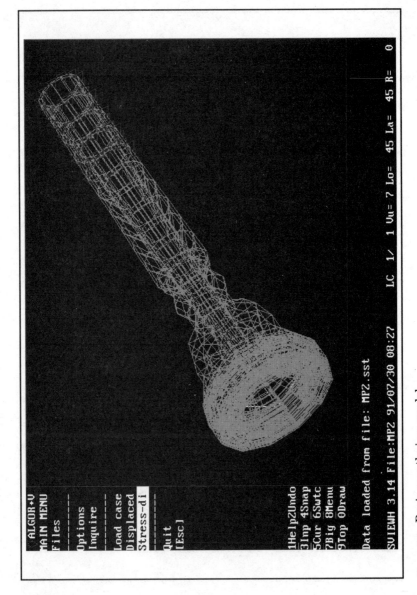

Figure 8.3 Basic mouthpiece model.

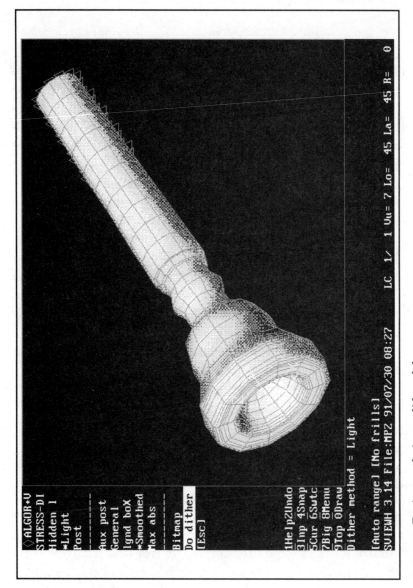

Figure 8.4 Basic mouthpiece solid model.

mode. Figure 8.5 shows the first mode bending; it is at 681 Hz. The second, third, and fourth mode frequencies are at 681 Hz, 2582 Hz, and 5358 Hz, respectively.

8.3 Modeling the Modified Mouthpiece

The modified mouthpiece consists of a brass ring approximately 0.75 inch in diameter by 0.98 inch in length. The inner diameter fits the taper of the shank of the mouthpiece. Figure 8.6 illustrates the profile before the full development of the part. Following Fig. 8.6 is Fig. 8.7, showing the three-dimensional finite element model.

Again, use is made of SuperView to render the model in solid perspective to check for any modeling defects. The solid model is shown in Fig. 8.8.

The natural frequencies for this model were 1162 Hz, 1162 Hz, 3461 Hz, and 4886 Hz. As can be seen, the additional piece of brass increased the natural frequencies by a factor of 1.71 for the first two modes, 1.34 for the third mode, and 0.91 for the fourth mode. The deformed shape for the first mode is given in Fig. 8.9.

It appears that one could keep the rim and inner dimensions of the mouthpiece and change the outside to reflect a heavy-walled design. The next problem addresses exactly this type of change.

8.4 A Modified Heavy-Wall Design Mouthpiece

The previous cross sections given above were modified to produce a heavy-walled design. All of the mouthpiece internals and rim are the same; only the outside from the mouthpiece/leadpipe junction to the rim was changed. This change gave a much heavier mouthpiece and consequently much higher natural frequencies. Figure 8.10 shows the cross section with the rotated pieces.

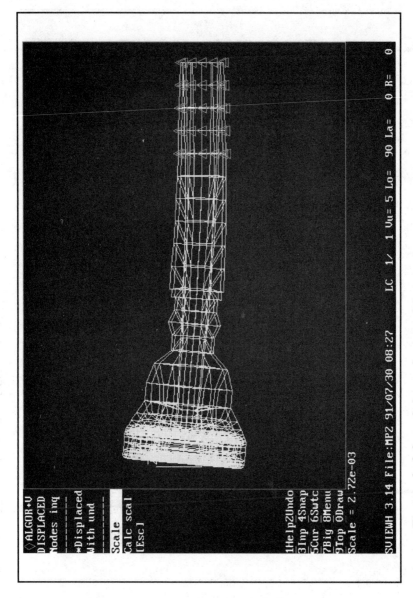

Figure 8.5 First mode deformed shape—basic model.

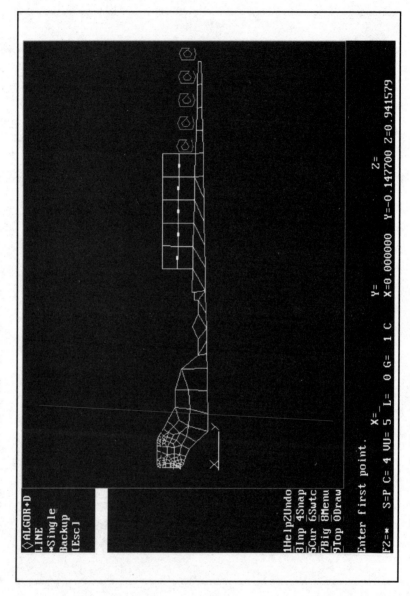

Figure 8.6 Basic model with intensifier ring—profile.

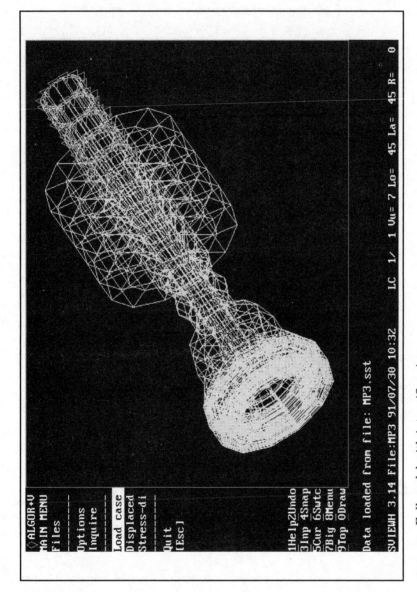

Figure 8.7 Full model with intensifier ring.

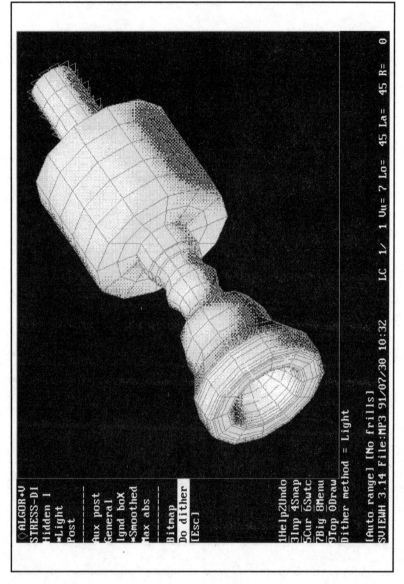

Figure 8.8 Solid model with intensifier ring.

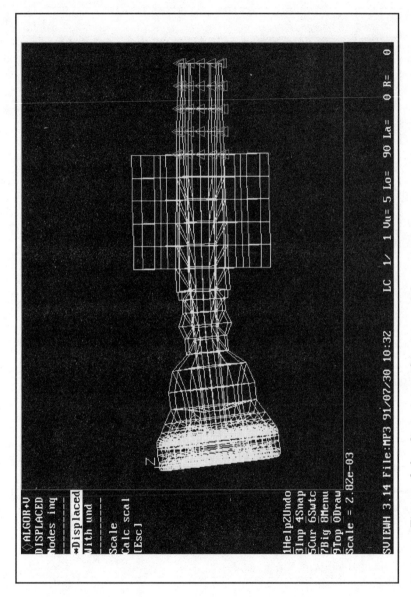

Figure 8.9 First mode bending—intensifier ring model.

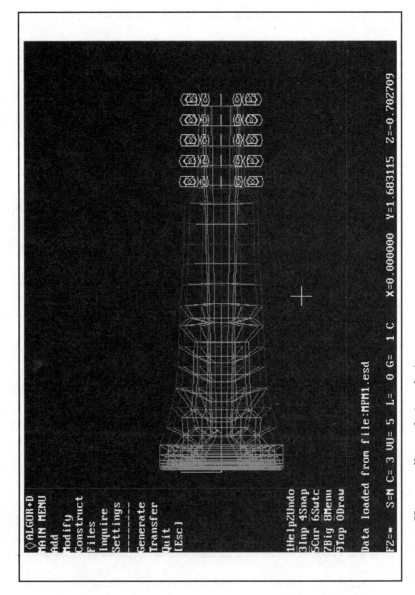

Figure 8.10   Heavy-wall mouthpiece design.

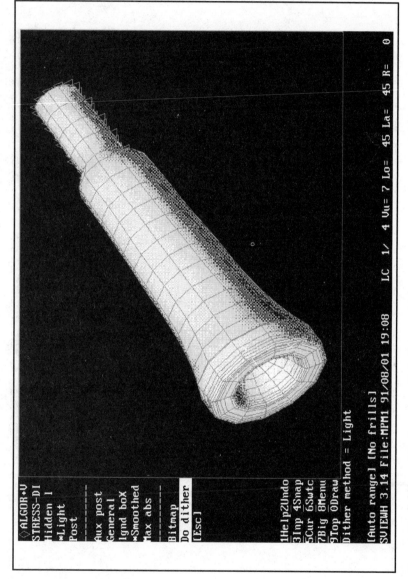

Figure 8.11 Solid model—heavy-wall design.

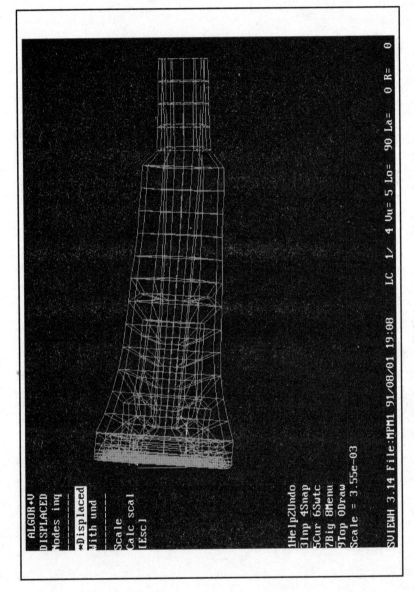

Figure 8.12 First mode bending—heavy-wall design.

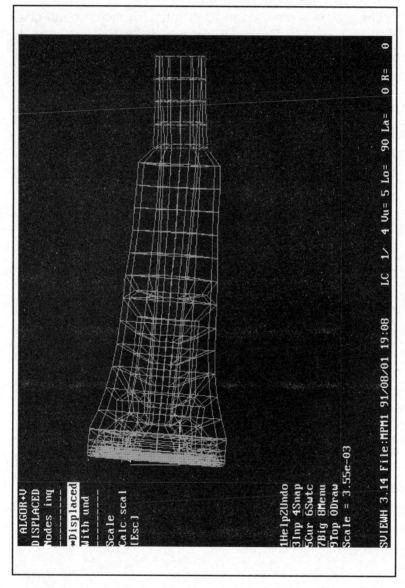

Figure 8.13 Fourth mode bending—heavy-wall design.

The mouthpiece is designed such that the large part starts just at the mouthpiece/leadpipe junction. Figure 8.11 shows the solid rendering of the model.

The analysis resulted in first to fourth mode frequencies of 1142 Hz, 1142 Hz, 4952 Hz, and 8778 Hz, respectively. Over the base model the increase in frequencies is by factors 1.68, 1.68, 1.92, and 1.64 for the four modes, respectively.

Figures 8.12 and 8.13 show the first and fourth bending modes of the modified mouthpiece.

Chapter

9

Applications to
Military Products

9.1　Minimizing the Product Design
Cycle by Interfacing CAD and FEA

With the decreasing amounts budgeted for defense oriented
programs, companies must be able to approach product
design from new perspectives and be assured of product
performance before the product is built. Companies must be
poised and ready to refine the existing designs and to make
innovative and sometimes bold use of new technologies.
One avenue that has been successful for many companies
has been the integration of PC-based CAD/CAE and mainframe CAD and PC-based CAE in the areas of electronic
design, optical design, and mechanical design. From the
first stages of a design, the analysis is an integral part of
the overall design effort. It has been found that preliminary
conceptual design modeling with the appropriately defined
environmental restrictions, loads, and forcing functions can
alleviate unforeseen problems in the final design stages by
determining the suitability of that design to perform to
specification. Problem areas are identified and immediately
cycled into the design process to resolve the problem areas
and to modify the design where required.

This section touches upon the area of mechanical design and how mechanical performance parameters are verified through use of Finite Element Analysis. A compressor/heatsink unit which is part of a LASER Rangefinder to be possibly used in the US Army M1A1 Abrams tank is examined using modal analysis and steady state heat transfer.

Unfortunately, not every engineer has at his disposal a large mainframe computer, the finite element code, and the time to solve whatever problem may arise. Consequently, the engineer is forced to make approximations based on similarity to classical, closed-form solvable problems, or the engineer may rely on physical prototyping. Physical prototyping is generally expensive and increases the cost of the design process. These approximations may or may not result in an adequate design. An overdesigned (or overengineered) product may be noncompetitive from a cost standpoint and/or may not actually perform as well as intended. With the proliferation of personal computers and the availability of software to aid the design process, it is no longer necessary to rely heavily on intuition and to settle for best guess solutions to real-world problems. FEA programs are available to aid the engineer in the development of a product to withstand its environment.

9.2   The Cryo/Heatsink Model

Figure 9.1 illustrates the compressor/heatsink of the LASER Rangefinder. The model was generated from CAD generated cross-sectional views.

Given the fixed design of the cryogenic compressor, a heatsink was designed that provided a means of holding the unit in place and providing an adequate thermal contact area for removal of heat from the compressor. Due to an extremely tight schedule, only preliminary estimates of heatsink performance and design were performed. Final performance estimates are presented in this section. Two options exist for translating the CAD-generated data into a useable finite element model. The first option is to transfer

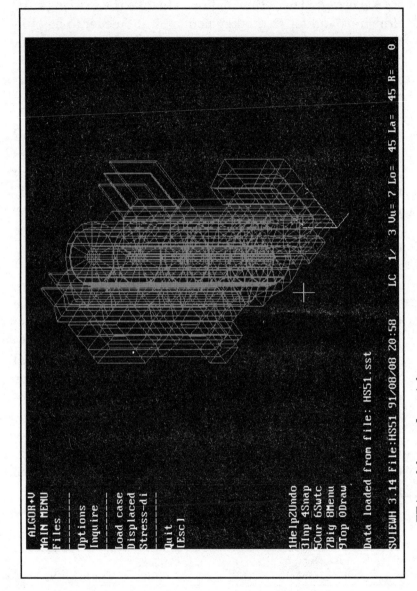

Figure 9.1 FEA model—cryo/heatsink.

239

the files directly to the finite element program with translators written by the user or provided as part of the CAD or FEA package. The second option is to take the existing CAD drawings and enter the key points of the model to be built into the program by the keyboard, digitizer, or mouse. For this particular problem, the second option was chosen primarily due to program demands on the CAD group during the time of this analysis.

In order to perform the analysis for this particular piece, the finite element package SuperSAP by ALGOR Interactive Systems, Inc. was chosen. This package provides a complete PC-based design tool. SuperSAP includes a front end three-dimensional CAD package developed specifically for the modeling and the decoding of models for use in the finite element program. In addition, the modeling package contains a three-dimensional graphic display and visualization program that allows viewing of deflected shapes, stress states, and general light shaded views for model inspection.

The physical properties are given in Table 9.1. Since the thermal gasketing material is compressed, the stiffness of the gasketing material was adjusted to account for the compression-induced stiffness. In a similar manner, a material was "created" that represented the weight and the stiffness of the engine. The remaining material, except for the potting compound, was A356 cast aluminum.

The unit shown is comprised of 800 nodes and 500 three-dimensional Isotropic Brick elements. All major features of the unit are incorporated into the model, except for the inclusion of the small fins opposite the mounting plane. Figure 9.2 is the solid rendering of the unit to be modeled.

TABLE 9.1   Material Properties for Cryo/Heatsink Problem

#	Ref temp	Weight density	E	μ	Expan coeff	Cond	Sp ht
1	70.0	0.0980	10.3E6	0.3	1.3E-5	7.1759E-4	0.111
2	70.0	0.0794	05.0E5	0.3	1.3E-5	7.1759E-4	0.111
3	70.0	0.0607	05.0E5	0.3	1.3E-5	7.1759E-4	0.111
4	70.0	0.1542	30.0E6	0.3	6.0E-6	7.1759E-4	0.111

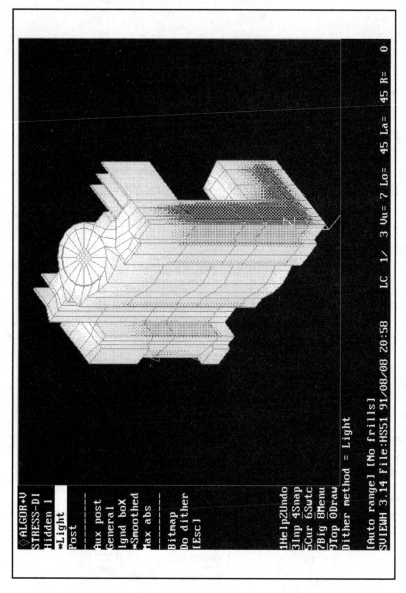

Figure 9.2 Solid rendering—cryo/heatsink model.

TABLE 9.2   First Three Fundamental Frequencies

Mode	Frequency (Hz)
1	805.79
2	838.63
3	1093.70

9.2.1   Analysis results of the cryo/heatsink model

A modal analysis was performed on the unit to determine the first three fundamental frequencies. These are given in Table 9.2.

These frequencies are approximately 550 Hz above the first resonance of the LRF, and the other frequencies do not coincide with the frequencies obtained from modal testing and analysis of the LRF. Figure 9.3 shows the first mode bending response of the unit.

This basic information is used to determine the response of the unit to other types of excitation. Table 9.3 gives the maximum displacements for the unit at resonance.

For example, consider the response of the compressor and

TABLE 9.3   Maximum Deflections (in) Modal Analysis

Global X Translation	Global Y Translation	Global Z Translation
NODE 806	805	682
3.0035E-03	1.2673E-03	1.3883E-04
NODE 678	682	539
3.0013E-03	1.2071E-03	1.2898E-04
NODE 805	699	683
2.9880E-03	1.2063E-03	1.2806E-04
NODE 807	685	684
2.9880E-03	1.2050E-03	1.1839E-04
NODE 677	694	540
2.9858E-03	1.2047E-03	1.1568E-04
NODE 679	707	420
2.9858E-03	1.2041E-03	1.1395E-04
NODE 808	724	685
5.1539E-04	1.2024E-03	1.1288E-04
NODE 809	807	688
5.1133E-04	1.2019E-03	1.1146E-04
NODE 680	715	542
4.9514E-04	1.2003E-03	1.0788E-04

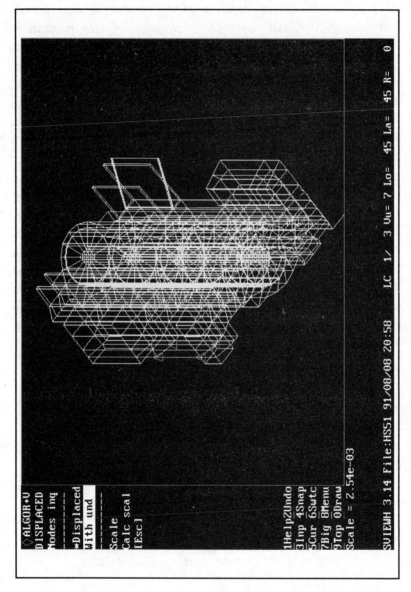

Figure 9.3   First mode deflection—cryo/heatsink model.

heatsink to a random vibration input. *Random vibration* is motion which displays no apparent pattern when recorded over time (the actual pattern of the motion is not repeatable); however, the frequency content of the motion is repeatable. In random vibration analysis, the excitations are assumed to follow a Gaussian distribution with zero mean value. Since this vibration is a random phenomenon and the method is based on probability theory, predictions can be made concerning the probability of exceeding a given displacement or stress. Table 9.4 gives the maximum deflections for a random vibration analysis performed on the unit.

The next area to examine in this design was the heat transfer performance of the system to see how the temperature levels in the structure are distributed. This problem was not difficult to solve since the model was built for the structural analysis. Differences in the two types of analysis were the use of a three-dimensional 8-node Thermal Brick Element, specifying the heat source, and specifying the proper convection coefficient from the appropriate surfaces.

TABLE 9.4   **Maximum Deflections (in) Random Vibration**

Global X Translation	Global Y Translation	Global Z Translation
NODE 808	809	809
3.7883E-04	2.3265E-05	1.0878E-05
NODE 680	805	680
3.7502E-04	2.3178E-05	1.0588E-05
NODE 809	807	681
3.7491E-04	2.2202E-05	1.0566E-05
NODE 681	808	808
3.7110E-04	2.2013E-05	1.0518E-05
NODE 806	804	679
3.4376E-04	2.1138E-05	9.9300E-06
NODE 678	803	805
3.4339E-04	2.0316E-05	9.9240E-06
NODE 807	682	677
3.4193E-04	1.9697E-05	9.8513E-06
NODE 805	699	807
3.4192E-04	1.9684E-05	9.8461E-06
NODE 679	685	678
3.4156E-04	1.9666E-05	9.8303E-06

Although the results presented herein are for a steady state analysis, a transient analysis would have been as straight-forward using the SuperSAP package. Thermal results are easily ported to the structural model for estimates of thermally induced stresses. Results of the thermal analysis are shown in Fig. 9.4 and the thermally induced stresses are shown in Fig. 9.5.

It has been shown in this section that it is possible to obtain reliable estimates of part-performance from an integrated PC-based FEA program, and that CAD software can be done quickly and with a smooth transition of data from one discipline to another. Personal computer-based analysis is a viable alternative to costly mainframe analysis in most instances. Extremely large models can be handled on a PC, provided you have large amounts of RAM, a large and fast hard disk, and a 386 platform. Platforms (33 MHz and faster 486 machines, etc.) will considerably enhance the PC alternative for analysis.

9.3   Analysis of a Tray for Aircraft Electronic Storage

This problem involves the analysis of a tray used inside a military aircraft that contains two electronic boxes each weighing approximately 36 pounds apiece. Initially, the tray was fabricated with 0.062-inch aluminum, reinforced with an aluminum bar across the rear of the tray unit. The tray unit plugged into its compartment via two dagger pins at the front (fan area) of the tray and was retained in place by being bolted to the aircraft interior. There are four fans that provide cooling for the electronic units. Initial tests and field trials revealed that the unit was not capable of prolonged use due to the development of cracks in the tray. The goal of the analysis was to minimize redesign and remanufacture of the tray and to make up time on an already late delivery schedule.

It was decided to utilize FEA to isolate the problem and

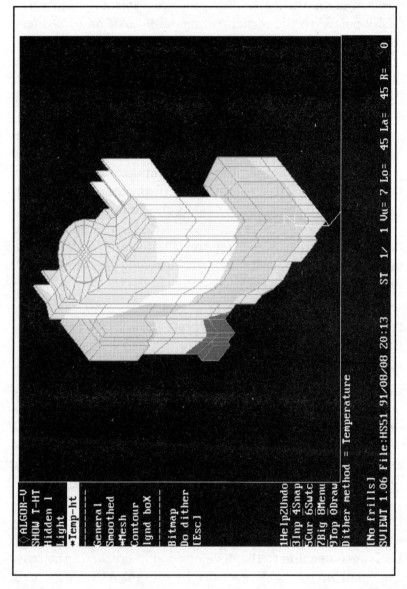

Figure 9.4 Thermal profiles—cryo/heatsink model.

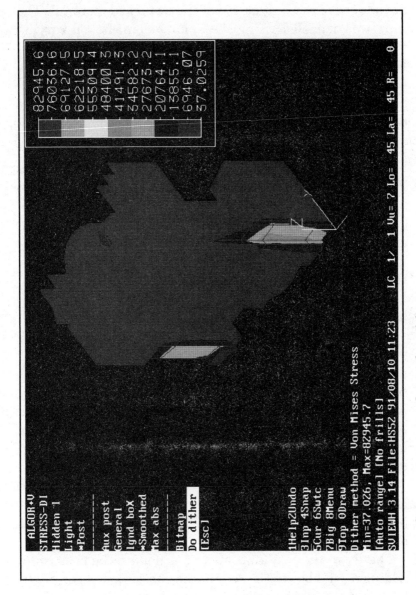

Figure 9.5 Stress contours—cryo/heatsink model.

determine how the cracks formed and where they first developed. Initial test results showed a tear in the material at the rear of the unit on top of the upturned material. The unit was modeled using 416 plate elements of varying thicknesses to develop a baseline for modification studies. Figure 9.6 shows the FEA model and Fig. 9.7 illustrates the solid rendering of the model.

9.3.1   Tray analysis results—not modified

The model was run to determine the first three modes of bending and the natural frequency associated with each. Frequencies were found at 238.91 Hz, 242.23 Hz, and 277.29 Hz. Stress levels were checked at each of the frequencies. The tray showed no abnormally high stress levels anywhere in the tray that would cause the observed cracking. Figures 9.8 through 9.10 show the bending modes for the first three frequencies.

Figure 9.11 shows the stress levels for the first mode bending.

Examination of the test unit showed that during vibration the right front tie-down point was broken. The dagger pin had worn into its receptacle and had allowed the right front edge of the unit to vibrate free. The resulting natural frequencies for the first three modes was calculated to be 96.18 Hz, 240.78 Hz, and 243.40 Hz. The resulting stress levels for the tray in the first mode are illustrated in Fig. 9.12 for a 1G sinusoidal input.

For the 5G input, the resulting stress level would be beyond the yield point of the material. The location for the break (highest stress value) corresponds exactly to that observed during test.

The fix to the tray consisted of stiffening the chassis by adding a solid bar to the rear of the chassis, solid rectangular bars to the chassis edges, and a member across the chassis width to tie the two mounting points (dagger pins) together. A view of the rear modified section is shown in Fig. 9.13.

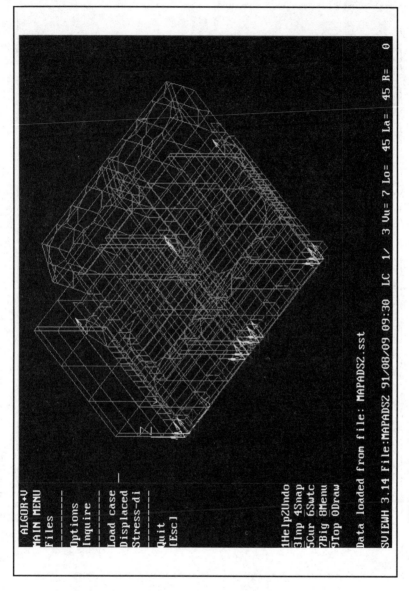

Figure 9.6 FEA model—electronic tray support.

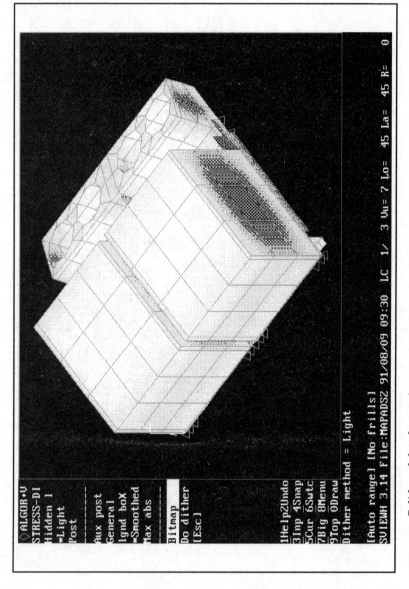

Figure 9.7 Solid model—electronic tray support.

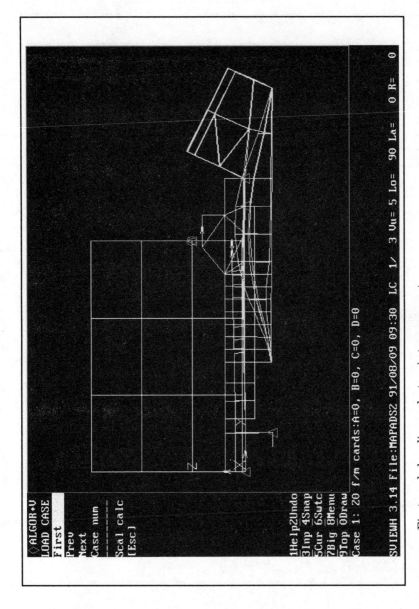

Figure 9.8 First mode bending—electronic tray support.

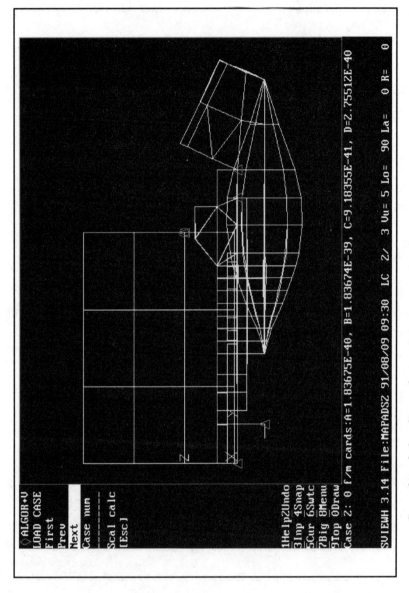

Figure 9.9 Second mode bending—electronic tray support.

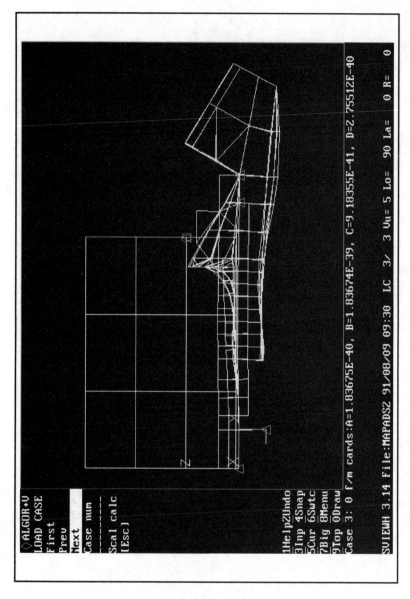

Figure 9.10 Third mode bending—electronic tray support.

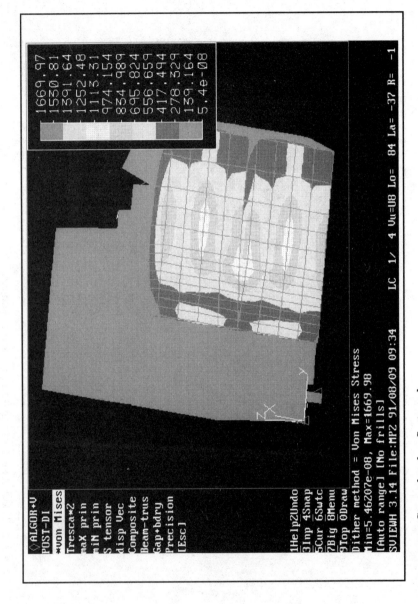

Figure 9.11 Stress levels—first mode.

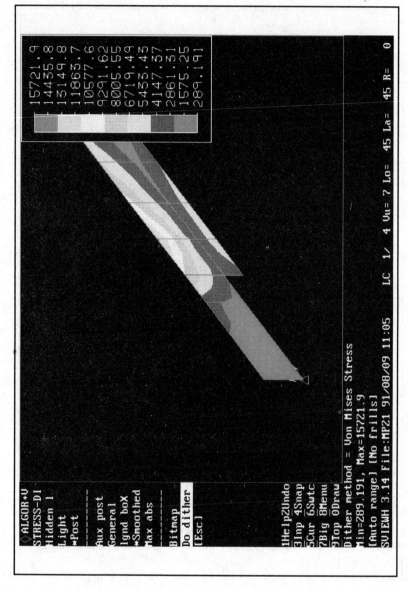

Figure 9.12 Stress levels—front tie-down removed.

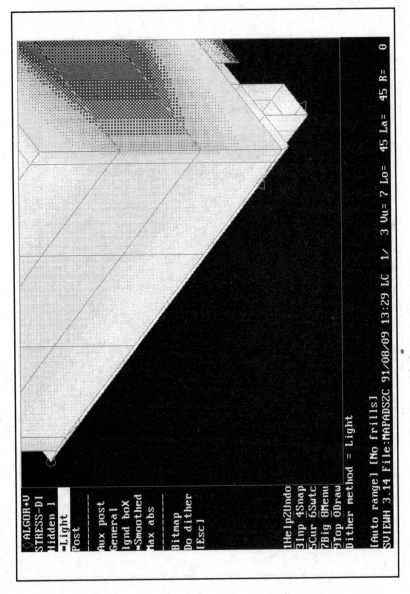

Figure 9.13 Modified tray showing added rear support.

With the added stiffness, the stress levels in the entire tray are reduced significantly and the area of concern, shown in Fig. 9.14, now has no stress-related problems. The mounting tray has passed all qualification tests and is now flying in the designated aircraft.

9.4   Production Support—FEA of Undersized Lenses

9.4.1   Summary of problem

Structural analyses were performed on an infrared zoom telescope to determine whether a production lot of under-sized lenses could be used without system degradation.[30] These lenses met the clear aperture and other optical requirements; however, it was necessary to verify that the resulting change in the lens bond thickness would not sig-nificantly increase the stress and distortion in the lens under severe environmental conditions. The fact that the telescope had originally been developed with the aid of finite element methods, and the math model had been vali-dated and maintained throughout the development cycle, greatly facilitated this analysis. The results demonstrated that there would be no structural degradation and the lenses could thus be salvaged.

9.4.2   Introduction to lens/housing stress problem

The subject telescope is an infrared zoom lens assembly. This unit is a militarized, E-O (electro-optic) system which provides both zoom and automatic athermalization capabil-ities for FLIR (forward looking infrared) systems[31]. The optics consist of stationary lenses plus two worm-driven lens cells supported on a pair of guide rails. The lens cells are positioned by external electrical commands for zooming and by an internal feedback control loop to accomplish the automatic athermalization. The optical assembly plus the

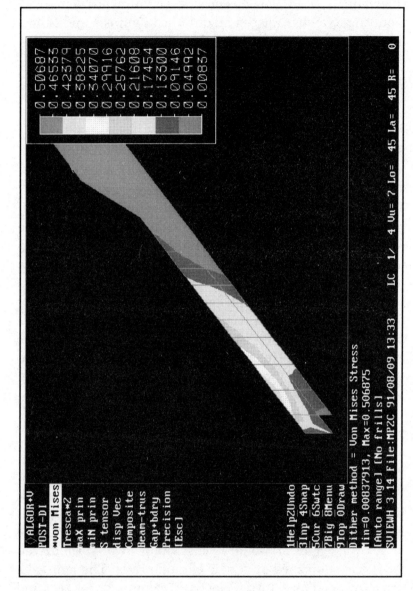

Figure 9.14 Stress concentration—modified tray.

electro-mechanical parts are packaged in a weathertight, A356-T6 cast aluminum housing. The unit meets stringent optical performance and boresight retention requirements under full military shock, vibration, and thermal stress environments with a minimum volume and weight.

The lenses in question were originally specified to be a nominal 1.781 inches in diameter, but a production lot of the Number 2 Lens was received with nominal 1.755 inch diameters. These lenses are fabricated from germanium (for infrared transmission), which is a relatively expensive and long lead material. In addition, by the time a finished lens is received, a considerable investment has already been made in cutting and polishing the blanks to the final optical prescription. Cost and schedule demands were thus the impetus for performing these analyses with the goal of salvaging these parts.

The difference in the specified versus the actual lens diameters would seem small, but the telescope is a precision E-O device where even tenths of a mil movement of optical elements can significantly degrade the system performance. For military use, this precise alignment must be maintained under the hostile shock, vibration, and thermal stress environments associated with worldwide use on multiple ground, air, and sea-going platforms. This is not a trivial task considering the nonconventional structural materials involved. The germanium optics, for instance, are brittle with low thermal expansion coefficients relative to the aluminum housing into which they are mounted. To accommodate these mismatches without overstressing and distorting the optics, the lenses are mounted with a thermal expansion compensating elastomer, RTV (room temperature vulcanizing), in a radial gap. This gap is carefully sized to minimize thermal stress while maintaining adequate support under shock and vibration loads. The undersized lens results in a deviation of this gap size from that specified, which could increase the stress and distortion in the lens. While the stresses are not likely to increase to the point of failure, permanent lens shifting could occur, and/or the system performance could become temperature-dependent due to lens thermal strain.

9.4.3 System level finite element analysis for E-O systems

During the original product development process, detailed engineering analyses were performed to meet the telescope weight and performance requirements. This included finite element shock, vibration, and thermal analyses. A parallel CAD/CAE/CAM product development procedure[32] was used in which a system level finite element model was created early in the development cycle and continuously updated as the design progressed. This procedure resulted in few major design iterations, a successful qualification test on the first attempt, and, as a side benefit, provided a validated math model of the system for production support and/or future product modifications. Having a validated system model available made it a simple matter to simulate the under-sized lens problem analytically.

System level finite element models are used for several purposes in E-O design[33]:

To predict average stresses for sizing the major structural members and initial material selection. Detail stress analyses of critical and/or high gradient areas are analyzed separately with a refined mesh using the boundary conditions from the system level analyses.

To predict the internal response levels for component design/selection and subcontractor specifications.

To predict the lens/mirror spacing shifts, decenter and tilt during shock, vibration, and thermal stress. These results are used to predict system performance (boresight, defocus).

The modeling goal at the system level is thus to get average stress calculations for major member sizing and accurate dynamic response predictions, while maintaining a manageable model size for design iterations and dynamic response analyses. For performance prediction and design verification, however, additional detail is required in the optics beyond what is normally expected in a system structural model. As a minimum, first order local bending of the lenses/mirrors is necessary for change in radii calculations. This requires five nodes per half wave length of the mode shape of interest. Also, 15° sectors are required for cylindrical surfaces to achieve acceptable accuracies with low order plate elements. With these considerations, an E-O finite element system model can very quickly grow to a cumbersome size, particularly for dynamic response calculations.

The system level finite element model of the telescope is shown in Fig. 9.15 for the external surfaces and Fig. 9.16 for the internal optics.

Consistent with system level modeling practices, the model is constructed primarily of beam and isoparametric, quadrilateral plate elements. Equivalent beam elements are used to simulate bearing, bolt, shaft, and elastomer bond stiffnesses. Lumped masses account for nonstructural parts such as motors, electronics, etc. The number and distribution of nodes in the model were optimized for computation speed with acceptable accuracy. About 2000 nodes were ultimately required.

In addition to utilizing the debugging aids supplied within the finite element preprocessor (ANSYS Prep7[34]), the finite element model was checked against as many physical measurements as was practical.

The static properties were first verified using static equivalent inertia loads to compare analytical mass/stiffness distributions versus independent weight/center of gravity estimates. The static analyses were also used for the initial material selection and structural member sizing.

The natural frequencies and associated mode shapes were then extracted to determine the dynamic properties of the system. These dynamic properties were later correlated

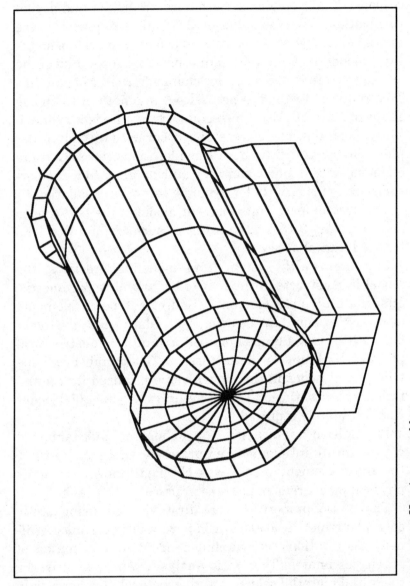

Figure 9.15 Housing assembly.

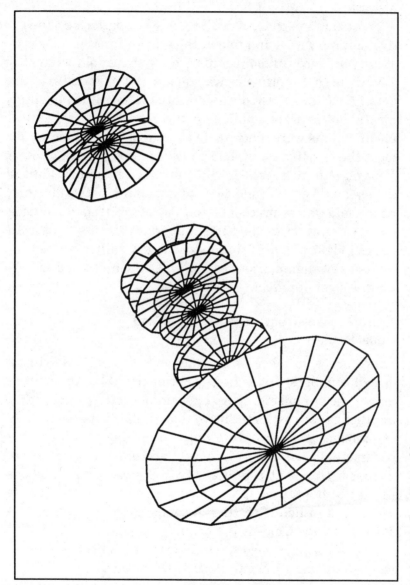

Figure 9.16 Lens assembly.

with the results of modal testing on actual hardware and adjustments in the finite element model were made where necessary.

Frequency response calculations were performed to verify damping ratios and to obtain transfer functions for sensitivity analyses and/or comparison with modal testing.

With a good mathematical model of the structure, the detail structural design was initiated. The critical environmental loads were applied to the model, and structural modifications were made to define a structure which would meet the specifications. The analyses were done in an iterative manner to approach a fully stressed structure. For the stress analysis, in addition to standard safety factors, microyield (a.k.a. precision yield) stress criteria were used for the optical parts and their support structures. This is an optics industry rule of thumb which generally assures that the optical alignment/boresight will be maintained following the stress producing event.[35]

9.4.4   Finite element model validation and modal testing

There are a number of circumstances where verifying a design completely by test may not be possible. An environment (nuclear effects, space applications, etc.) can be either impractical or cost prohibitive due to elaborate test set up or damage to a costly test unit. In optical systems, some parameters (e.g., lens responses) are difficult to measure during a transient condition. In these cases, an analysis using a validated computer model must be used to qualify the design in lieu of testing. A comparison of analysis versus test mode shapes is the primary means of validation.

Modal testing is used synergistically with FEA (finite element analysis).[36] Some structural properties, such as damping and interface boundary conditions, are more easily extracted from test data than predicted analytically. Conversely, important system parameters, such as optical deflections during transients, are difficult to measure in the laboratory but readily available from the computer analyses. The computer analyses also help during the modal testing to identify frequency ranges of interest and coincident, multiple natural frequencies. These may not be found during test if the test engineer does not suspect their existence.

One of the advantages of modal testing is that a complete assembly is not necessarily required for useful correlations studies. Prototype parts and/or subcomponents of the system, which may be available early in the development cycle, can be tested and correlated to similarly configured finite element models. That was the case for this telescope, where several prototype housings and subassemblies were fabricated for developmental tests. The housings were originally developed using the Pro-Engineer solid modeling system.[37] Here the parts were reviewed for producibility, then transferred electronically to CAM to lay out the tool paths and create the NC (numerical control) machine tapes. The prototype housings were machined from these tapes without having to create formal drawings, and were thus available for testing very early in the program.

As an example, the outer housing, a critical structure providing environmental enclosure and objective lens support, was prototyped and used for partial finite element validation.[38] The housing was tested using a Bruel & Kjaer Type 2034 Dual Channel FFT Signal Analyzer[39] in conjunction with PCMAP Modal Analysis software.[40] The housing was tested in a free-free state by simply suspending it from rubber bands. An accelerometer was fixed on the housing at a point which, based on the FEA, would respond during the first five modes of vibration. An impact hammer/force transducer was used to excite the structure at various points on the housing and calculate the associated transfer functions between the accelerometer and force transducer with the FFT signal analyzer. The transfer functions were then curve fitted using the PCMAP software, and the eigenvalues, eigenvectors, and residues were calculated. The mode shapes were then compared to the ANSYS predictions for a similarly configured finite element model of the housing for verification. The first mode shape is shown in Fig. 9.17.

The specified input vibration frequency ranged from 5 to 500 Hz. The specified shock pulse width was 40 msec which, in terms of an equivalent frequency, also falls within the 5 to 500 Hz band. With a one octave margin, the frequency range of interest for test and analysis was, thus, up to 1000 Hz. The major modes considered within this frequency range are summarized in Table 9.5. Excellent agreement was achieved for the lower modes with decreasing accuracy with higher frequency (as expected).

After achieving reasonable agreement between test and analysis for the modes in the frequency range of interest, the finite element model of the structure was considered validated. The detail structural design and weight optimization was then completed, leaving a validated finite element model available to support production and possible future design modifications.

Structural analyses were performed on the telescope by modifying the diameter (to 1.755 inch) of the Number 2 Lens (the subject undersized element) in the finite element model, and comparing the results to the previous analyses with the originally specified diameter (1.781 inch). From the previous analyses, it was known that the design drivers were a low temperature soak requirement of $-46°C$ (the worst case thermal stress environment) and random vibration along the optical axes. The random vibration require-

TABLE 9.5   Modal Correlation Study (Housing)

Frequencies Are in Hz.

Mode number	Test freq	FEA freq	Diff
1	562	546	−03%
2	740	760	+03%
3	820	848	+03%
4	972	1071	+10%
5	1420	1312	−08%

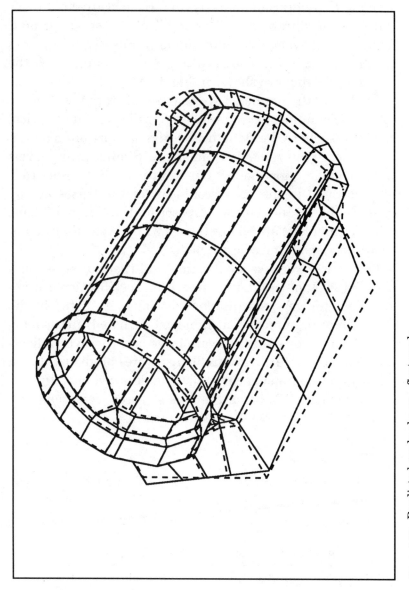

Figure 9.17 Predicted mode shape—first mode.

ment consists of multiple military platforms specified from *MIL-STD-810D*, (figures 514.3-10 through 514.3-22).[41] To simplify the vibration analysis (in a conservative manner), the random vibration profiles of all of the specified platforms were enveloped into a single profile. This profile is shown in Fig. 9.18. The resulting stress contours on the external housing are shown in Fig. 9.19.

As shown, the predicted stresses are all very low relative to A356-T6 aluminum allowables. This is typical of optical structures where the structural design is more often driven by stiffness requirements than stress. Similar stress levels were calculated for the low temperature (−46°C) soak case as illustrated in the stress contours for the lenses in Fig. 9.20. As shown, the thermal stresses for the Number 2 Lens (second from left) are so low (below 21 psi) that they do not even show up on the contour plot.

The actual worst case vibration and thermal stress values for the Number 2 Lens, one with a specification compliant diameter versus an undersized lens, are listed in the stress analysis summary in Table 9.6. As listed, all the margins of safety for either case are large and positive, indicating that no structural problems were predicted for a telescope system using the subject undersized lens. In fact, except for the RTV bond, the lens stresses were generally reduced due to the smaller lens diameter.

9.4.5 Conclusions

Based on the structural analysis performed here, it was concluded that the undersized lenses (1.755 inch diameter) could be used in the telescope with no degradation in struc-

TABLE 9.6 Stress Analysis Summary Number 2 Lens

Load case:	−46°C		Random vib		Min margin of safety
Material:	Germanium	RTV	Germanium	RTV	
1.781 Lens	31 psi	−3 psi	608 psi	53 psi	+0.31
1.755 Lens	18 psi	−6 psi	396 psi	51 psi	+1.02

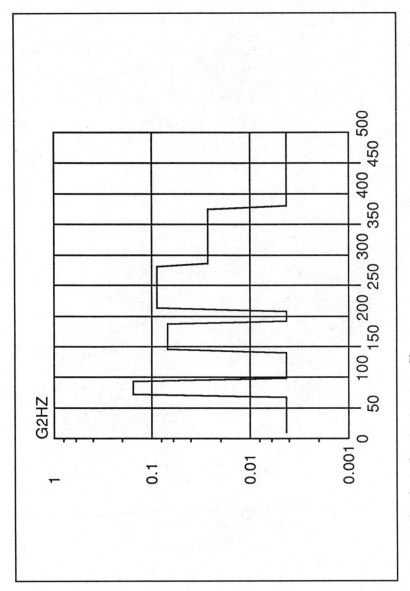

Figure 9.18 Random vibration input profile.

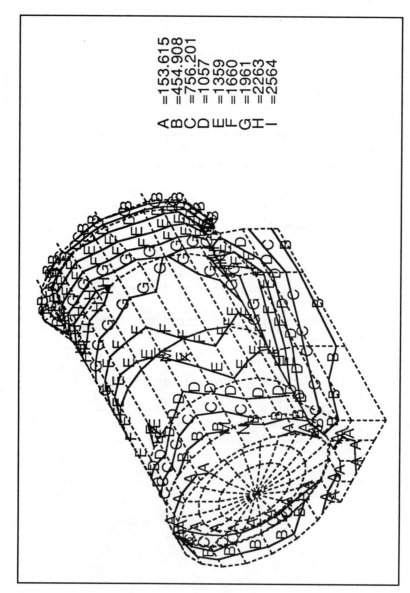

A =153.615
B =454.908
C =756.201
D =1057
E =1359
F =1660
G =1961
H =2263
I =2564

Figure 9.19   Random vibration stress levels.

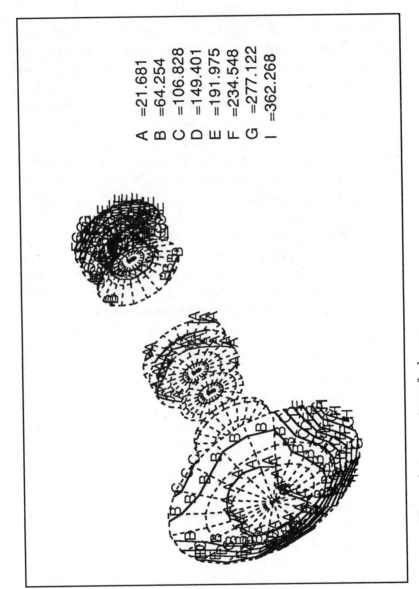

A = 21.681
B = 64.254
C = 106.828
D = 149.401
E = 191.975
F = 234.548
G = 277.122
I = 362.268

Figure 9.20 Thermal stress contours for lenses.

tural performance. Subsequent thermal and vibrational acceptance testing, along with field experience, have confirmed the analytical predictions.

This work also demonstrated the usefulness of maintaining a validated system level finite element model, not only in the development cycle, but throughout the production life cycle as well. Access to the validated finite element model greatly facilitated production support, in that production problems could be simulated analytically in a prompt and cost effective manner. In this case, a relatively expensive, long lead-time piece of hardware was shown to be usable with no impact on system performance. In another case, the results could have been a prediction that the parts should not be used, thus avoiding costly field returns of defective systems. For the finite element model to be useful for this purpose, however, it must be emphasized that the math model must be properly maintained to represent product changes, and that it also must be thoroughly validated via modal or other suitable testing.

10

FEA in Consumer
Medical Products

10.1 Introduction

The area of consumer medical products is interesting from the standpoint that the manufacturer not only has to produce products that the consumer will desire, but also products that the Food and Drug Administration will have to sanction for public use. In addition, there are agencies that have to certify the unit for safe consumer use. Examples of this are the Underwriters Laboratories and the Canadian Standards Association.

The engineer is faced not only with satisfying the above, but also ensuring that the product is assembled in a minimum amount of time using piece parts that just meet passable criteria. The bottom line is product margin and therefore company profitability. Adding to the overall profitability picture is the amount of time and development effort that can be placed in a product. Tight schedules often lead to design oversights that must be solved after the product is in the tooling phase or, worse, in the production phase.

This chapter looks at three problems: two that deal with certain design aspects of a passover humidifier, and one that looks at a damping ring for a motor/blower assembly and the attempts to alleviate balance problems in the overall system. These problems will provide insight into the use of the ALGOR fluids module and an introduction to the nonlinear module within the ALGOR system of modules.

10.2   The Nozzle Problem

This problem arose from the design of a passover humidifier. An air stream is blown into a covered reservoir of water. As the air travels from the inlet to the exit of the device, the air's moisture content is increased. Measurements indicate that one can expect approximately a thirty percent increase in moisture content. The increase in moisture depends upon the residence time in the humidifier and state of the air flow.

The first part of the problem is concerned with the nozzle itself. In the initial design, as the air stream entered the humidifier it was noted that the air did not flow from the nozzle smoothly through the humidifier and out of the exit. Instead, visual observations showed that a large portion of the water area behind the nozzle appeared to have a standing eddy current. In addition, humidity levels at the exit were not as had been expected.

Figure 10.1 shows a cross section of the nozzle that was modeled initially. Note that this effort is only a two-dimensional one but is quite adequate for seeing where adjustments in the design needed to be made. It was assumed that the lower boundary, the water surface, remained stationary. In reality, as time progresses, the surface of the water will obtain a velocity due to the shear at the water-air interface and by the impact of the air stream into the water. The air stream does not initially flow parallel to the water surface. At some inlet velocities, the air stream will impact the water at an angle. Figure 10.2 illustrates the velocity field followed by the pressure field in Fig. 10.3.

To adjust the nozzle design to guide the fluid through the humidifier and over the water, the modification shown in

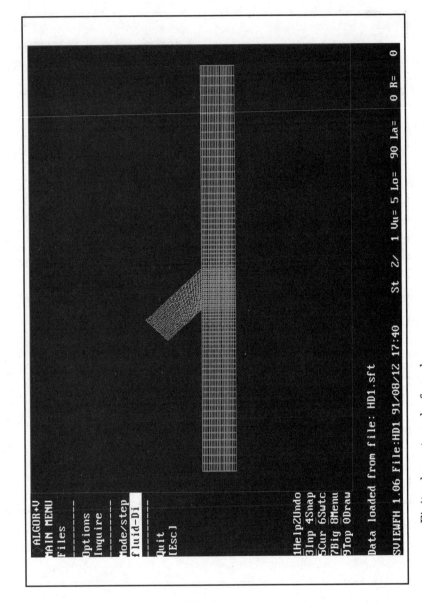

Figure 10.1 Finite element mesh of nozzle.

275

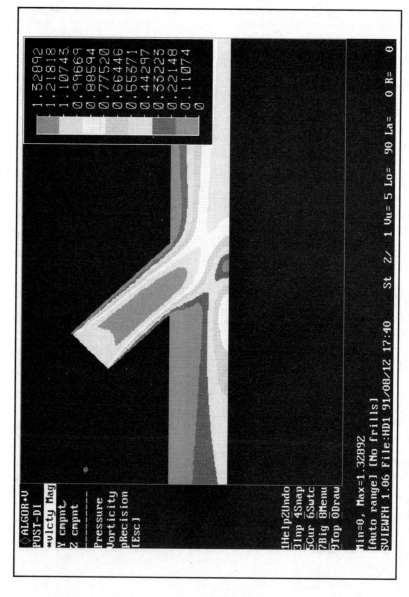

Figure 10.2 Velocity distribution in unmodified nozzle.

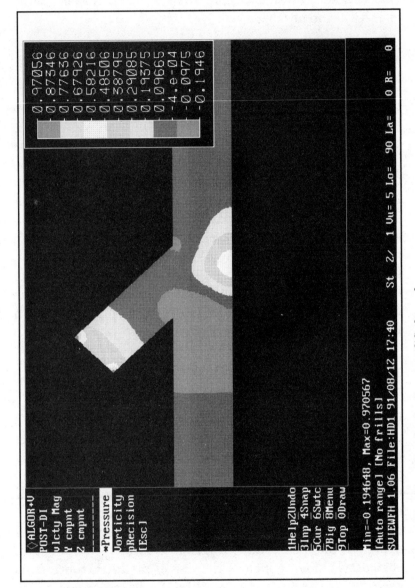

Figure 10.3 Pressure distribution in unmodified nozzle.

Fig. 10.4 was made to the inlet nozzle. Again, the analysis was run and the results are depicted in Figs. 10.5 and 10.6. One can see the much improved flow distribution in the vicinity of the nozzle.

10.3 The Flow Chamber Problem

Once the inlet nozzle problem was corrected, the problem that had to be addressed was that of internal flow within the humidifier. Figure 10.7 shows the plan view of the humidifier with the inlet and outlet ports. Many iterations on vane layout to direct the flow over the water and provide the longest residence time within the humidifier were studied. Finally, the configuration shown in Fig. 10.8 was chosen. This view shows the vane configuration and the finite element mesh. The mesh was generated with the automatic mesh generation capability of SuperDraw II. In addition, the middle vane was extended through the top of the humidifier to use as a handle in opening and closing the unit. The velocity field and the pressure field are shown in Figs. 10.9 and 10.10, respectively.

10.4 Analysis of a Vibration Isolator for an Impeller Mount Assembly

Obstructive sleep apnea is a condition where a person's upper airway collapses while sleeping, causing cessation of breathing. The person awakens enough to begin breathing and falls back to sleep. Usually this process occurs many times during the night, with the person being unaware of it ever happening. Periods of interrupted breathing make a person drowsy during the day and can be physically damaging to the heart and lungs. A common treatment of obstructive sleep apnea is the use of a CPAP (continuous positive airway pressure) machine to act as an air splint and hold the upper airway open.

The problem discussed involves an aluminum centrifugal impeller mounted to a brushless DC motor that is part of a

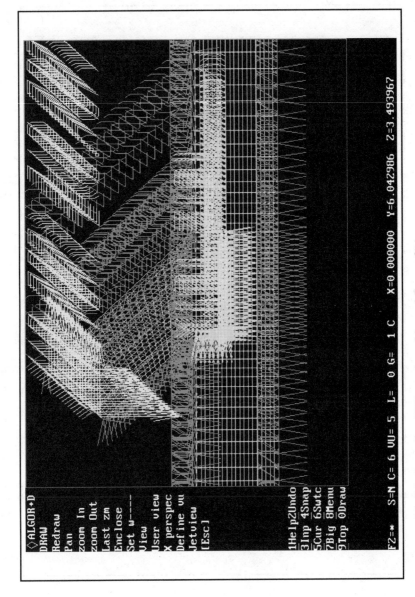

Figure 10.4 Modified nozzle—FEA mesh and boundary conditions.

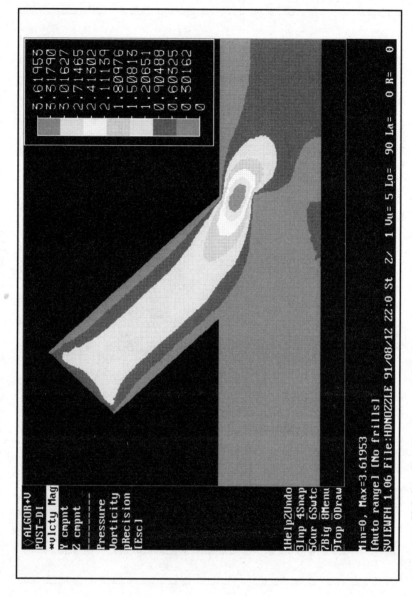

Figure 10.5 Velocity distribution—modified nozzle.

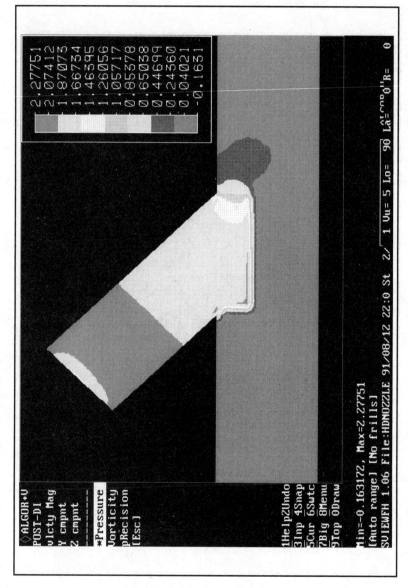

Figure 10.6 Pressure distribution—modified nozzle.

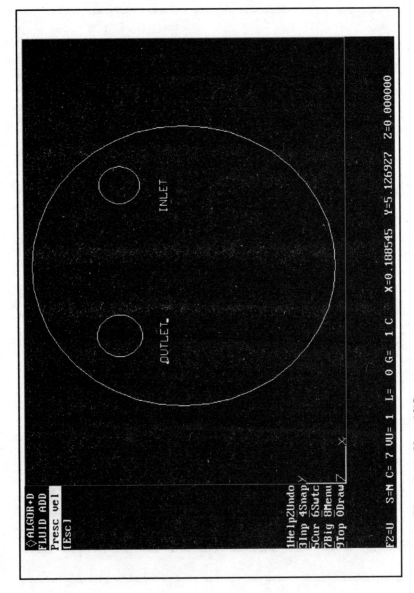

Figure 10.7 Plan view of humidifier.

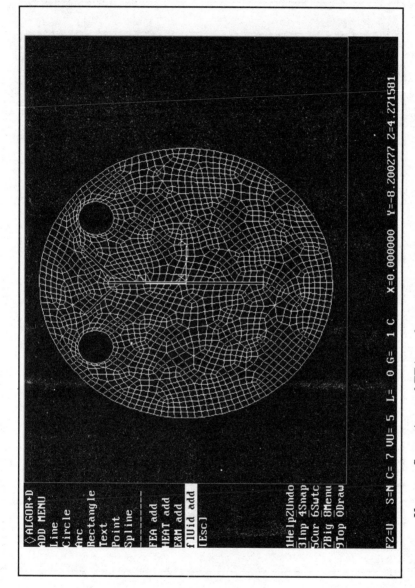

Figure 10.8 Vane configuration and FEA mesh.

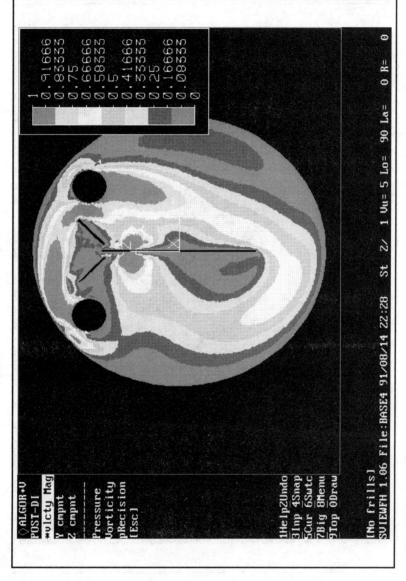

Figure 10.9 Velocity field in humidifier.

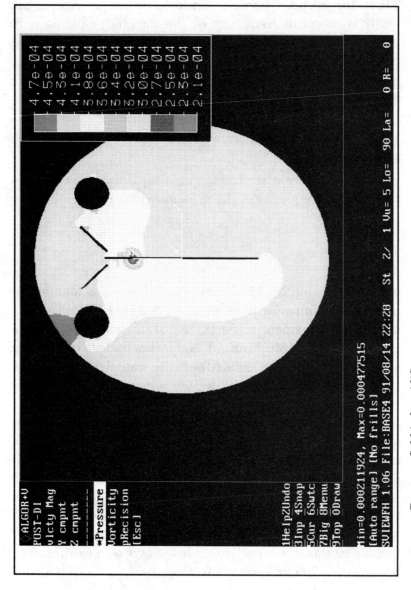

Figure 10·10 Pressure field in humidifier.

blower in a CPAP machine.[42] An inherent characteristic of the motor is that it develops torque pulsations as it rotates. These torque pulsations, referred to as torque ripple, can excite a resonant frequency in the impeller, causing it to emit a considerable amount of noise. The CPAP machine is used at home while a patient is sleeping, and must be as quiet as possible. For this reason, it is desirable to reduce the effect of torque ripple on the impeller.

10.4.1   Equations of the physical phenomena

Resonance occurs when the forcing frequency, w_f, is equal to the natural frequency of the system, which is given by

$$\omega_n = \sqrt{\frac{k_\theta}{J}} \qquad (10.1)$$

where k_θ is angular spring rate and J is mass moment of inertia. The impeller resonates at 1200 Hz (2400p rad/sec) and its mass moment of inertia is 7.3E-5 kg·m². Using these values, k_θ is 4150 N·m/rad. Experiments using rubber o-rings to isolate the motor from the impeller determined that a spring rate less than 100 N·m/rad is sufficient to reduce noise to an acceptable level.

The impeller mount must hold the impeller perpendicular to the motor shaft, in addition to isolating motor torque ripple. In order for the mount to hold the impeller perpendicular to the motor shaft, it must be rigid axially; but to isolate torque ripple it must be flexible in an angular direction. Several designs were considered, but the one that best fulfilled both of these requirements was a circular ring of flexible beams. Each individual beam acts as a cantilevered beam with a linear spring rate given by

$$k_l = \frac{3EI}{l^3} = \frac{Ewh^3}{4l^3} \qquad (10.2)$$

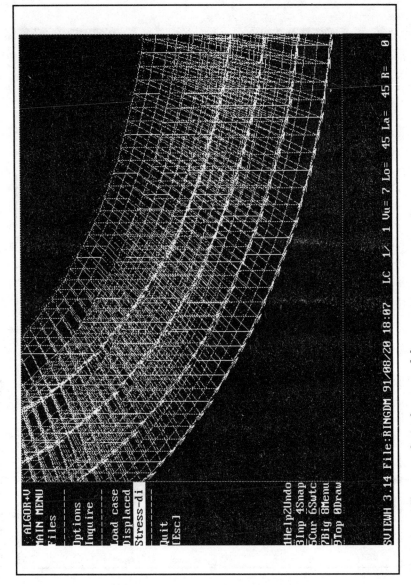

Figure 10.11 Close-up of wireframe model.

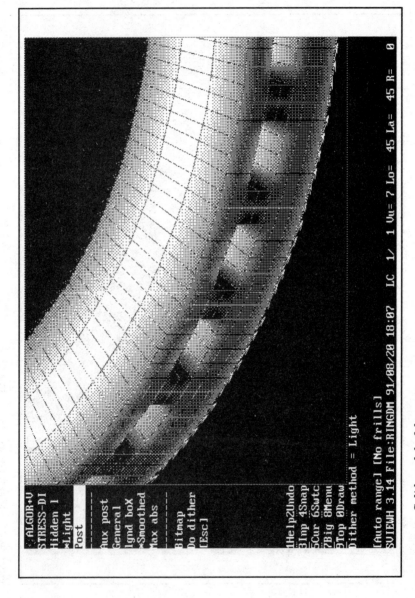

Figure 10.12 Solid model of damper.

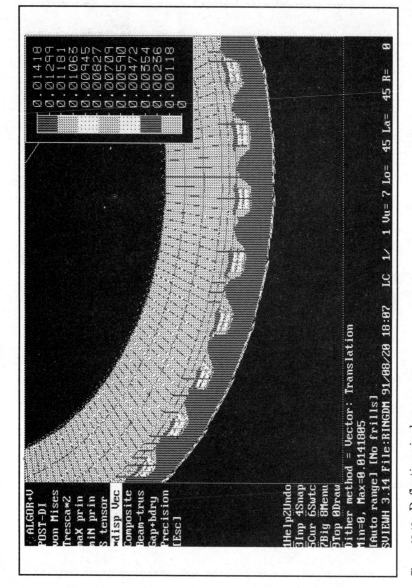

Figure 10.13 Deflections in damper.

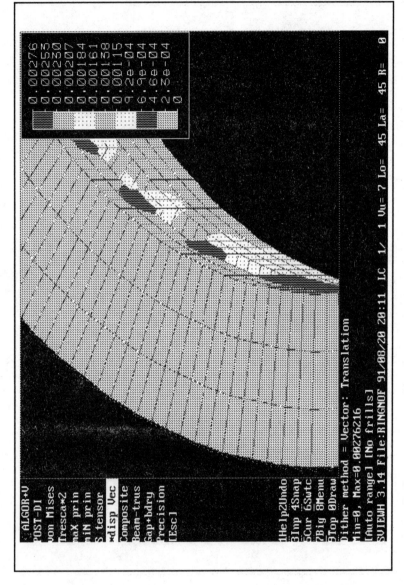

Figure 10.14 Deflections—vertical loading.

where E is the modulus of elasticity, w is the beam width, h is the beam height, and l is the beam length. Each individual beam of the circular ring has a relatively large value of w to give axial strength, and a relatively small value of h to give a low spring rate. The angular spring rate of the entire ring can be expressed as

$$k_\theta = \frac{T}{\theta} = \frac{Fr}{\theta} \qquad (10.3)$$

where T is torque, q is the displaced angle, F is force, and r is the radius for the applied force.

10.4.2 The FEA model

The model was constructed such that the upper surface was constrained and the lower surface was attached to a rigid ring to allow uniform movement in an angular direction. The wireframe model is shown in Fig. 10.11 and the solid model is shown in Fig. 10.12.

An arbitrary force was applied to the bottom of the rigid ring. The force was varied from outside to inside of the ring to assure that constant torque was being applied. The displacement of the ring is used to determine the angle of movement. This angle can be used in the equation for angular spring rate to determine the overall spring rate of the circular ring. The total force applied to the ring was 43.8 N at a mean radius of 0.02 m. The results of the analysis showed that the rotation angle was 0.018 rad, which gives an angular spring rate of 48.7 N·m/rad. The model was later subjected to an upward force of 4.4 N on one side and a downward force of 4.4 N on the other to simulate the effects of gyroscopic torque from the impeller. The results of this analysis showed that the compression of the rubber ring was 0.0027 mm, which is sufficient to prevent the impeller from contacting the blower housing. The FEA model has shown that a molded rubber ring consisting of many small beams is sufficient to isolate torque ripple from the motor, as well as to provide enough stability to the assembly to prevent the impeller from contacting the blower housing.

11

An Example of FEA in the Utility Industry

11.1   Introduction to Transformer Pad Analysis

This section contains the results of a Finite Element Analysis of the Electri-Glass BP2000 Fiberglass Transformer Box Pad. The purpose of the analysis is to verify the product performance under loading conditions as found in the Electri-Glass Structural Test Report on the BP2000 dated January 17, 1981.[43]

The analysis used was a linear static stress. Since plastics and fiberglass materials do not behave linearly where the load-deformation behavior is accurately described by the classical theory of elasticity, the FEA model was run with a loading of 1000 lb_f. The test data used for comparison was normalized to inches of deflection per 1000 lb_f load.

Test results indicated that deflections in the Box Pad were at 0.062 in per 1000 lb_f load with a ± value on the deflection measurement of 0.032 in. Analysis results indicated that deflections in the Box Pad ranged from 0.0015 in to 0.113 in. These deflection numbers indicate that for the linear portion of the load-deflection curve for this material application, the linear static FEA model performs well.

11.2 The Finite Element Model

Figure 11.1 shows the FE Model used in this analysis. Figure 11.2 is the solid rendering of Fig. 11.1. The model is composed of 568 three-dimensional plate elements using the Hiesh, Clough, and Tocher type plate bending elements. Each quadrilateral shell element consists of an assemblage of four triangular elements. The stiffness from each subtriangle is transformed into the local mean plane of the quadrilateral element. The excessive degrees of freedom are statically condensed out. This does not restrict the four plate nodes to be coplanar; therefore, the warpage in the shell structure can be properly modeled.

In addition, the modeling software contains a feature that produces the appropriate stress values at the points or nodes of interest rather than the mathematically modeled Gaussian points used in some FEA programs. This provides for more accurate results for the model.

All material values were taken from correspondence from Electri-Glass and are as follows:

Flexural strength	16e3 psi
Flexural modulus	10e5 psi
Tensile strength	9e3 psi
Tensile modulus	8e5 psi
Compressive strength	15e3 psi
Elongation	1%
IZOD impact strength	4 ft·lb/in of notch
Sp gr	1.72

The walls were modeled with 0.1875-in thickness while the top was modeled with 0.25-in thick plates. External ribbing per details from Electri-Glass was included, as well as molded in wooden stacking blocks. Model weight from the FEA program was 92+ lb.

Loading was simulated as described in the test report referenced above. The loading was maintained at 1000 lb$_f$ to

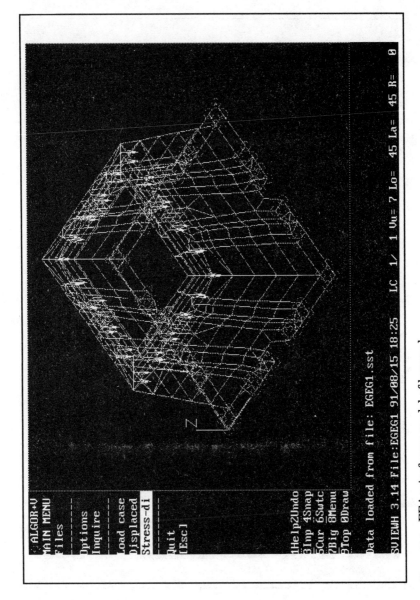

Figure 11.1 FEA wireframe model of box pad.

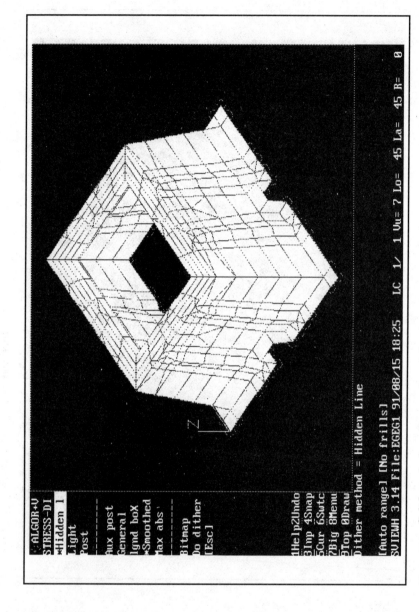

Figure 11.2 Solid rendering—box pad model.

remain in the linear region of the stress-strain curve of the material.

Nonlinear analysis involves the load-deformation interaction. Deformation or displacements within the model change the way the load interacts with the structure, which in turn affects future deformation. Nonlinear analysis constantly takes into account the possible deformations and deflections, and feeds this structural information back into the system during the analysis. Since the nature of the nonlinear stress-strain relationship of the fiberglass material used in this product was not known, the assumption of a linear stress-strain material property was used to estimate deflections and stresses for the first 1000 lb_f of applied loading.

The model was assumed to be resting on a nondeflecting base such that the only model restraints consisted of allowing no deflection in the $-z$ (gravity) direction.

11.3   Results of the FEA Analysis

Results of an FEA are best presented in terms of stress plots and deflection illustrations. Since the validity of the modeling effort was based on the ability to predict the actual deflections of the unit, I have elected to present these results first.

Figure 11.3 indicates that deflections range from 0.0015 in to 0.113 in. The maximum deflections are higher around the inner edge of the Box Pad opening and gradually taper to deflections on the order of 0.01 in. In general, deflections around the vertical walls show deflections of 0.01 in to 0.076 in. Maximum deflections occurred approximately 6 to 8 in below the top surface. It appears that one driver for the sidewall deflections may be the molded-in stacking blocks. As the top surface lip deflects downward and tends to rotate inward toward the vertical wall, the loading is partially transferred to the sidewall via the blocks.

Stress levels were not severe for this loading, the maximum being approximately 1275 psi as shown in Fig. 11.4.

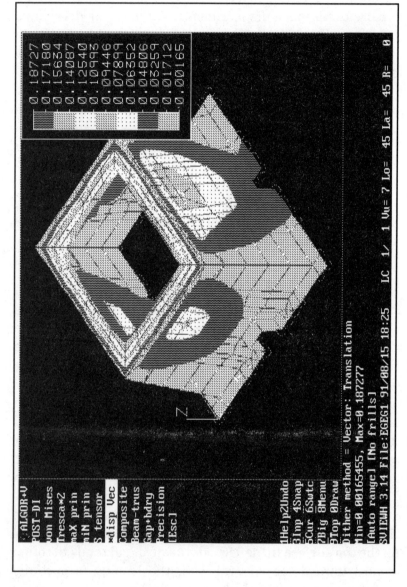

Figure 11.3   Deflection of box pad model.

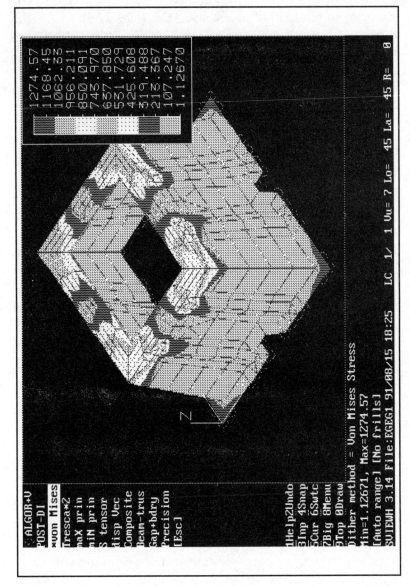

Figure 11.4 Stress levels in box pad model.

Measured results indicate that at a loading of 12,000 lb$_f$ to 14,000 lb$_f$ for a Box Pad weighing 92+ lb, the Box Pad ruptured. If we assume a linear stress-strain curve for the material, a twelvefold increase in the load would push the stress levels to 15,000 psi or greater. This value is less than the flexural strength and above the tensile strength of the material. However, any Box Pad less than perfect in material and form could see failure at less than the published nominal value of the material strength.

11.4   Conclusions on Modeling of the Box Pad

Test results indicated that deflections in the Box Pad were at 0.062 in per 1000 lb$_f$ load with a ± value on the deflection measurement of 0.032 in. Analysis results indicated that deflections in the Box Pad ranged from 0.0015 to 0.113 in. These deflection numbers indicate that for the linear portion of the load-deflection curve for this material application, the linear static FEA model performs well. Additional modeling of similar items and comparison to test data indicate that this type of modeling is good. Using this approach, it is possible to examine design modifications to existing parts using the linear region of the stress-strain curves for the material. For large deflections and/or material nonlinearities, it is better to use the nonlinear static module of the ALGOR software.

There are differences in the analysis results and the measured test data. I believe that these differences can be accounted for in the following ways:

1. Actual measurements are accurate to ±0.032 in. Therefore, the deflections could range from 0.032 to 0.096 in, which is more consistent with the predicted values.

2. Proper accounting for material nonlinear behavior may have yielded somewhat more accurate answers. The nonlinear aspects could account for some redistribution of the loading vector during deformation.

Appendix

ALGOR Interactive Systems, Inc. has been developing and marketing engineering software for mechanical engineers since 1977. Among the many products offered by ALGOR is the ALGOR Finite Element System.

ALGOR began as a time-sharing service that provided analytical solutions on mainframe computers to subscribing engineers. After developing special-purpose engineering software for a number of subscribers, ALGOR entered the FEA market by offering FEA programs through its time-share service. The company created a number of customized interface programs that eased FEA data entry and processing and provided customized plotting and output data processing. In 1983, ALGOR decided to take the FEA material offered at that time for Prime computers and rewrite the code for the PC. In October of 1984, ALGOR delivered the first full-featured FEA package for the PC/XT. Since there were no programmed-in limitations on model size, ALGOR was an immediate alternative to mainframes.

Today, ALGOR offers a wide variety of programs to handle many types of analysis problems. The following paragraphs address the major modules. At the end of the Appendix, the company address and phone number are given so that you may contact them for further information and current pricing.

ViziCad Plus—Modeling and Design Visualization for FEA

The ViziCad package includes the program SuperDraw II for modeling an entity. It is a completely graphical program for creating models of designs that are to be analyzed. With this program, the user is able to add boundary conditions, boundary elements, nodal temperatures, nodal forces, and moments. The second element of the package is the Decode module that translates the graphical format into the format required for the type of analysis being undertaken. Within this module, the user can create custom material libraries. The decoder also creates input for the next module of the package, SuperView.

SuperView provides the complete environment for displaying and examining the analytical results. SuperView also includes an option for precision contouring that allows the user to determine the relative accuracy of the model. The precision contouring highlights areas in the model where refining the mesh could produce more accurate results.

The ViziCad program supports most common graphics boards and contains a number of CAD interchange formats. These are:

Autocad (9)	DXF	2-D
Autocad (10)	DXF	3-D and points
Versacad	TWG	2-D
Cadkey	CDL	3-D and points
CV Personal Design	SGX	Imports from .DAT file
Anvil 1000	NFL	2-D
Micro Cadam	DXF	3-D and points
Triumph	CDL	3-D
MicroStation	DXF	2-D
IGES	IGS	3-D

Stress and Dynamic Analysis

The Stress and Dynamic Analysis package analyzes mechanical designs under many different types of loadings.

A variety of processors are included to analyze static and dynamic loading situations. These are:

Linear static stress

Weight and center of gravity

Mass moment of inertia

Dynamic modal analysis

Time history analysis using modal superposition

Response spectrum analysis using modal superposition

Time history analysis using direct integration

Buckling analysis of Beam element models

Dynamic analysis with load stiffening of Beam element models

Steady-State and Transient Heat Transfer

In some designs, the mechanical forces may produce minimal stress while the thermal effects can drive the design critical. The heat transfer package contains two separate modules, one for steady-state analysis and the other for transient heat transfer analysis. The thermal processors allow the user to understand how a model will respond to thermal effects.

Nonlinear Stress and Dynamic Analysis

Although most designs remain within the linear elastic region, many complex designs can experience deformations and material behavior that extend into the nonlinear region. Because the materials and geometry of some designs lead to material yielding and large deformation, many industries, such as the automotive industry (in examining crash-worthiness of cars), are turning toward nonlinear analysis.

Composite Analysis

For stress and dynamic analysis of composite materials, ALGOR provides the processors. The user assigns laminate orientation, and the package automatically generates the complicated composite properties.

Other Modules

ALGOR provides modules for the analysis of problems involving the following:

Random Vibration

Buckling

Modal Analysis with Load Stiffening

Nonlinear Gap/Cable Element Analysis

Frequency Response Analysis

Beam Design Editor

Kinematic/Rigid Body Dynamic Analysis

Iconnex M.E. Workbench

PipePlus Pipe Stress Analysis

Electrostatic Analysis

Fluid Flow Analysis

There are other add-on programs for modeling enhancement. The address and telephone number for information are:

ALGOR Interactive Systems, Inc.

260 Alpha Drive
Pittsburgh, PA 15283
(412) 967-2700
Fax: (404) 967-2781

References

1. Clough, R.W., "The Finite Element Method in Plane Stress Analysis," *Proceedings of 2nd ASCE Conference on Electronic Computation,* Pittsburgh, PA, September 8 and 9, 1960.
2. Courant, R., "Variational Methods for the Solutions of Problems of Equilibrium and Vibrations," *Bull. Am. Math. Soc.,* Vol. 49, 1943.
3. Morse, P.M. and Feshback, H., *Methods of Theoretical Physics,* McGraw-Hill Book Company, New York, 1953, Section 9.4.
4. Greenstadt, J., "On the Reduction of Continuous Problems to Discrete Form," *IBM L. Res. Dev.,* Vol. 3, 1959.
5. Turner, M.J., Clough, R.W., Martin, H.C., and Topp, L.C., "Stiffness and Deflection Analysis of Complex Structures," *J. Aeronautical Sciences,* Vol. 23, No. 9, 1956.
6. Melosh, R.J., "Basis for the Derivation of Matrices for the Direct Stiffness Method," *AIAA J.,* Vol 1, 1963.
7. Jones, R.E., "A Generalization of the Direct-Stiffness Method of Structural Analysis," *AIAA J.,* Vol. 2, 1964.
8. McLay, R.W., "Completeness and Convergence Properties of Finite Element Displacement Functions—A General Treatment," AIAA 5th Aerospace Science Meeting, New York, 1967.
9. Johnson, M.W. and McLay, R.W., "Convergence of the Finite Element Method in the Theory of Elasticity," *J. Appl. Mech.,* Vol. 35, No. 2, June 1968.
10. Tong, P. and Pian, T.H.H., "The Convergence of the Finite Element Method in Solving Linear Elastic Problems," *Int. J. Solids Struct.,* Vol. 3, 1967.
11. Zienkiewicz, O.C. and Cheung, Y.K., "Finite Elements in the Solution of Field Problems," *Engineer,* Vol. 220, 1965.
12. Girault, V. and Raviart, P.A., "Finite Element Approximation of the Navier-Stokes Equations," *Lecture Notes in Mathematics,* 749 pp., 1979, New York, Springer-Verlag.
13. Teman, R., *Theoretical Studies On The Finite Element Method Applied To The Navier-Stokes Equations,* 1977, North Holland, Amsterdam.
14. Fortin, M., "Old and New Finite Elements for Incompressible Flow," *Int. J. Numerical Methods Fluids,* 1981, pp. 347–364.
15. Griffith, D.F., "An Approximating Divergence-free 9-node Velocity Element For Incompressible Flows," *Int. J. Numerical Methods Fluids,* pp. 323–346.
16. Thomasset, F., *Implementation of Finite Element Methods for Navier-Stokes Equations,* 1981, Springer-Verlag, New York.
17. Heywood, J.G. and Rannacher, R. "Finite Element Approximations of the Non-Stationary Navier-Stokes Problem, Part I: Regularity of Solutions and Second Order Spatial Discretisations," *SIAM J. Num. Analysis,* 1982.
18. Cullen, M.J.P., "Analysis and Experiments with some Low Order Finite Element Schemes for the Navier-Stokes Equations," *J. Computational Physics,* 1982.

19. Chung, T.J., *Finite Element Analysis in Fluid Dynamics,* 1978, McGraw-Hill, 378 pp.
20. Gallagher, R.H., *Finite Element Analysis Fundamentals,* Prentice-Hall, Inc., Englewood Cliffs, NJ, 420 pp.
21. Segerlind, Larry J., *Applied Finite Element Analysis,* John Wiley & Sons, New York, 1976, 422 pp.
22. Zienkiewicz, O.C. and R.L. Taylor, *The Finite Element Method Vol 1, Fourth Edition.* McGraw-Hill, New York, 1989.
23. Roark, R.J. and Young, *Formulas For Stress and Strain, Fifth Edition,* McGraw-Hill, New York, 1988.
24. Kardestuncer, H. (Editor), *Finite Element Handbook,* McGraw-Hill, New York, 1987.
25. Baker, A.J., *Finite Element Computational Fluid Mechanics,* Hemisphere/ McGraw-Hill, New York, 1983.
26. Strang, G. and G.F. Fix, *An Analysis of the Finite Element Method,* Prentice-Hall, Englewood Cliffs, NJ, 1973.
27. Lawson, H., Moore Special Tool Company, Inc., Personal Communication.
28. Wilson, C., Warn Industries, Personal Communication.
29. Levering, P., Webb Wheel Products, Personal Communication.
30. Woodard, K., Kollsman, Personal Communication.
31. "FEA Process Shaves Costs," *Design News* December 4, 1989, pp. 83–87.
32. Gray, L.B. and K.S. Woodard, "An Integrated Mechanical Analysis of an Electro-Optical System in a DoD Development Environment," *1989 ANSYS Conference Proceedings,* Swanson Analysis Systems, Inc., Pittsburgh, PA, 1989, Vol. 1, pp. 3.23–3.34.
33. Woodard, K.S., "On Using Ballistic Shock Test Data for Direct Input to Finite Element Models of Optical Systems," *Proceedings of the 60th Shock and Vibration Symposium,* David Taylor Research Center, Portsmouth, VA, 1989, Vol. II, pp. 255–263.
34. Swanson Analysis Systems, Inc., Houston, PA.
35. Marschall, C.S. and R.E. Maringer, *Dimensional Instability: An Introduction,* Pergamon Press, Oxford, 1977.
36. Rodamaker, M.C., "Integrating Modal Testing and Finite Element Models," *Sound and Vibration,* pp. 4–5, Jan., 1989.
37. Parametric Technologies, Inc., Waltham, MA.
38. Woodard, K.S. and L.B. Gray, "Finite Element Analysis of a Military Zoom Lens," *Computers in Mechanical Engineering,* ASME, 1990, Vol. 2, pp. 51–56.
39. Bruel & Kjaer Instruments, Inc., Marlborough, MA.
40. Vibration Engineering Consultants, Inc., Woburn, MA.
41. *MIL-STD-810D Environmental Test Methods and Engineering Guidelines.*
42. Zeigler, M., Healthdyne Technologies, Personal Communication.
43. Peters, W., Electri-Glass, Inc., Personal Communication.

Further Reading

Bathe, K.J., *Finite Element Procedures in Engineering Analysis,* Prentice-Hall, Englewood Cliffs, NJ, 1982.

Bear, F.P. and Johnson, Jr., E.R., *Vector Mechanics for Engineers, Statics and Dynamics,* McGraw-Hill Book Co., Inc., New York, 1962.

Becker, E.B., G.F. Carey, and J.T. Oden, *Finite Elements: An Introduction,* Vol 1, Prentice-Hall, Englewood Cliffs, NJ.

Desai, C.S., *Elementary Finite Element Method,* Prentice-Hall, Englewood Cliffs, NJ, 1979.

Martin, H.C. and G.F. Carey, *Introduction to Finite Element Analysis,* Prentice-Hall, Englewood Cliffs, NJ, 1973.

Index

ABOUT THE AUTHOR

Edward R. Champion, Jr., Ph.D., is a consultant and registered Professional Engineer in the areas of fluid/heat transfer analysis and structural analysis for a wide variety of commercial and military applications. This interest in the application of numerical methods to solve many types of problems has led to the development of several PC-based analysis programs for both industry and business. Dr. Champion is the author/co-author of several books (including *Finite Element Analysis* and *Numerical Analysis*) and articles (covering subjects such as fabric combustion phenomena and waste heat transfer). He is a member of the ASME.